ARCHAEOLOGICAL SURVEY OF INDIA, NEW IMPERIAL SERIES, VOL. XL

THE
ASTRONOMICAL OBSERVATORIES OF JAI SINGH

ARCHAEOLOGICAL SURVEY OF INDIA, NEW IMPERIAL SERIES, VOL. XL

THE ASTRONOMICAL OBSERVATORIES OF JAI SINGH

BY

G. R. KAYE

Fellow of the Royal Astronomical Society
Honorary Correspondent of the Archaeological
Department of India

GYAN

Published by

Gyan Publishing House
5, Ansari Road
Daryaganj, New Delhi-110002
Phone: 011-47034999, 9811692060
E-mail: books@gyanbooks.com

Distribution Network
gyanbooks.com
India, USA, Canada, UK, Australia, France

ISBN : 978-81-212-2634-9 (HB)
First Published, 1917

2nd Impression 2021

Printed at: Gyan Press, Delhi.

The
Astronomical Observatories of Jai Singh
Author: G. R. Kaye

THE

ASTRONOMICAL OBSERVATORIES OF JAI SINGH

BY

G. R. KAYE,

Fellow of the Royal Astronomical Society.
Honorary correspondent of the Archæological Department of India.

CALCUTTA
SUPERINTENDENT GOVERNMENT PRINTING, INDIA
1917.

PREFACE.

THROUGH the kindness of Sir John Marshall, C.I.E., Litt.D., etc., Director General of Archæology, and the Hon'ble Mr. H. Sharp, C.S.I., C.I.E., etc., Educational Commissioner with the Government of India, I was able in December and January 1915-16 to visit the old observatories at Delhi, Jaipur, Ujjain and Benares, and this volume is a direct result of my tour.

In the following chapters an attempt has been made to exhibit the known facts relating to Jai Singh's astronomical work and to describe his instruments in some detail. The subject matter conveniently arranges itself into the following divisions: (i) Jai Singh's preparation for his astronomical researches; (ii) his own publications; (iii) the instruments of his predecessors that he employed; (iv) the instruments he devised; (v) the observatories he built; and (vi) an estimate of his work, etc. The sixth section presented some difficulty—chiefly because of the somewhat erroneous idea that prevails that an account of Jai Singh is necessarily intimately connected with the history of Hindu Astronomy. To form a proper estimate of the value of Jai Singh's work and to place it in correct relationship with that of his predecessors it is, of course, necessary to have knowledge of the history of the development of astronomy before his time; and while there is abundance of literature on European and Muslim astronomy, there is at present no systematic account of Hindu astronomy generally available; so a second part of this work containing a fairly full account of Hindu astronomy was under contemplation. But this would have altered the character of the book and Jai Singh would have ceased to be its principal feature: also an account of Hindu astronomy will appear at the same time as this volume in the 'Open Court Classics of Science and Philosophy.' I have, therefore, here given only a summary of Hindu astronomy in so far as it is related to Jai Singh's labours and for further details refer to my other book.

This volume is primarily a tour report for the Archæological Department and therefore principally descriptive. That it leaves much to be accomplished, is to be regretted but it was inevitable; and, indeed, to attempt to make such a record perfectly complete would mean the indefinite postponement of its publication. I must therefore plead for some lenience of judgment and I trust that the intrinsic interest of the facts recorded will, in some measure, compensate for the inadequacy of the presentation.

It is now my pleasant duty to record my grateful thanks for help and encouragement. To the Durbars of Jaipur and Gwalior I am greatly indebted for their kind hospitality and for the valuable assistance given by their officers; and my thanks are specially due to Lala Chuni Lal, the Darogha Imarat, and Professor V. V. Tamhankar and Pandit Kedar Nāth of Jaipur; to Rai Bahadur Munshi Bal Mukan, the Sar Suba, and Pandit Sham

Sundar Sharma, Tehsildar, Ujjain; to Pandit Gokal Chand of Jaipur for his assistance at Benares and Pandit Mahedeva Śāstri Ghatri who, with the kind permission of Her Highness the Maharani of Dharbanga, placed himself at my disposal at Benares; and to Mr. Sohan Lal of the Archæological Department who accompanied me on my tour. To Mr. Fazil Elahi, B.A., and Professor Abd-ur-Rahman of St. Stephen's College, Delhi, my thanks are due for assistance in translating some Persian works on the Astrolabe and in the interpretation of obscure Arabic terms. To the Superintendents of the Museums at Calcutta and Lahore I am indebted for the loan of certain instruments. The Public Works Staff of the Imperial City, Delhi, and particularly Mr. Glen, Executive Engineer rendered most valuable assistance; and to the care and skill of the Superintendent of Government Printing and the Surveyor General and their staffs I am obviously greatly indebted. It is impossible to repay in words my debt to Sir John Marshall and Mr. W. E. Jardine, C.I.E., the Resident of Gwalior, for advice, encouragement and help.

G. R. KAYE.

November , 1917.

CONTENTS.

CHAPTER IX.—UJJAIN OBSERVATORY.

CHAPTER X.—BENARES OBSERVATORY.

CHAPTER XI.—MATHURA OBSERVATORY.

CHAPTER XII.—HISTORICAL PERSPECTIVE.

CHAPTER XIII.—THE EVOLUTION OF JAI SINGH'S INSTRUMENTS.

CHAPTER XIV.—CONCLUSION.

APPENDICES.

LIST OF ILLUSTRATIONS.

The Astronomical Observatories of Jai Singh

Chapter I.—JAI SINGH.

1. Mahārāja Sawāī Jai Singh II of Jaipur was born in A. D. 1686[1] and succeeded to the Amber territory at the age of thirteen in A.D. 1699, a few years before the death of Aurangzīb. He had difficulties in establishing himself, but in 1708 obtained complete possession of the province. In 1719 he was appointed by Muhammad Shāh governor of the province of Agra and soon after to Malwā. In 1734 he was again governor of Malwā and in that year, apparently with the cognizance of the Emperor, he resigned the province to the Peshwā. He died in 1743, two hundred years after Copernicus, and " his wives, concubines, and *science* expired with him on his funeral pyre."[2]

Jai Singh mixed in most of the trouble and warfare of the long period of anarchy that coincided with his reign; but he distinguished himself more as a statesman than a soldier and has been termed the Machiavelli of his day. He was the founder of a new capital, named after him Jainagar or Jaipur, which in his time became a centre of learning; he erected caravanserais in many of the provinces; and he built astronomical observatories at five of the principal cities of Hindustan. He conceived and carried out a scheme of scientific research that is still a notable example; and his influence is still a living one. The observatories he erected are, in the words of his historian, " monuments that irradiate a dark period of Indian History."[3]

At an early age Jai Singh showed a predilection for astronomical work and, according to his own account, by constant study he obtained a thorough knowledge of its principles and rules. He found the astronomical tables in use defective and set himself the task of preparing new ones. With this purpose in view Jai Singh took every means to ensure success. He attached himself to no particular school but studied Hindu, Muslim and European methods impartially. He collected astronomical books and had certain of them translated; he organised

[1] The year in which Newton's *Principia* was completed.

[2] *Annals and Antiquities of Rajast'han.* By Lieutenant-Colonel James Tod, 1829, Vol. ii, p. 308.

[3] Tod, ii., p. 360.

a regular staff of workers and sent some of them to foreign countries to collect information; he invited certain Europeans and others interested in astronomy to Jaipur; he built a large observatory at Delhi and made careful observations there for seven years with a view to the preparation of a new star catalogue; and afterwards he built other observatories at Jaipur, Ujjain, Benares and Mathurā. Such in brief were his astronomical activites which we now proceed to describe in some detail.

Astronomical Works consulted by Jai Singh.

2. Of the works of his predecessors and contemporaries there is evidence that Jai Singh was acquainted with the following: Ptolemy's *Almagest*; the astronomical tables of Ulugh Beg; some Treatises on the Astrolabe; La Hire's *Tabulae Astronomicae*; and Flamsteed's *Historia Cœlestis Britannica*; also certain western mathematical works such as Euclid's *Elements*, a treatise on plane and spherical trigonometry and on the construction of logarithms. This, of course, cannot be an exhaustive list since his valuable library no longer exists entire,[1] and it would be fairly safe to assume that Jai Singh collected and studied all the available astronomical works; indeed it is recorded specifically that he procured from Europe, besides the tables of La Hire, those of earlier dates.

(*i*) The book that held sway in Europe for a thousand years after its publication and among the Arabs for a thousand years after its translation was Ptolemy's *Almagest*. No other text-book that has ever been written had such a reputation. Jai Singh himself speaks of Ptolemy as one of the greatest astronomers and one of his most important acts was to order a translation from the Arabic of Ptolemy's great work. This translation, apart from its intrinsic value, has a somewhat special interest: its title *Samrāṭ Siddhānta* 'the supreme text-book' has practically the same meaning as the Arabic title *Al-Majisṭī* 'the greatest' and as the Greek title *μεγάλη σύνταξις* 'the great compilation.'[2]

According to Garrett[3] the *Samrāṭ Siddhānta* expresses Jai Singh's views on astronomy and this, probably, is quite true; but the implication that it was an original work composed by or for Jai Singh is wrong. It was written by Jagganāth one of Jai Singh's assistants who was quite unambiguous on its origin. He wrote:

Granthaṁ siddhānta samrājaṁ samrāṭ rachayati sphuṭam |
Tushṭtyai Srījayasiṁhasya Jagannāthāhvayaḥ kṛitī ||
Arabī bhāshayā grantho Mijāstīnāmakaḥ sthitaḥ |
Gaṇakānāṁ subodhāya gīrvāṇya prakṭī kṛitah. ||

Jagannāth's introduction contains, besides the usual invocation, (*a*) Eulogies of Jai Singh with which are references to certain events of some importance[4]

---

[1] It is said that Jaggat Singh gave Jai Singh's unrivalled library to a courtesan: it was thus despoiled and its treasures distributed among her base relatives." This would account for the meagreness of the information now available; but the tale does not altogether bear the impress of truth.

[2] Another title is *Siddhānta Sāra Kaustubha*. See AUFRECHT, *Cat. Sans. Man. Trin. Coll., Dublin. p. 75.*

[3] *The Jaipur Observatory and its Builder.* By Lieutenant A. ff. Garrett, R.E., assisted by Pundit Chandradhar Guleri. 1902. pp. 19 and 21.

[4] The Govind image episode, the *Vāja Peya* sacrifice, and the abolition of a certain tax (? the *Jezeya*).

(*b*) a list of instruments and (*c*) an explanation of the source from which the work was obtained.

Jai Singh, Jagannāth says, was clever in exhibiting the new methods with globes and other instruments; and that, with the help of certain learned mathematicians and astronomers, he had made observations of the stars. The instruments proper to an observatory are said to be (1) Naḍi Yantra (sun-dial), (2) Gola Yantra (Sphere), (3) Digaṁśa Yantra (Azimuth instrument), (4) Dakhshiṇo Digbhitt (Mural quadrant), (5) Vṛitta Shashtāṁśaka (An arc of sixty degrees placed in the meridian) which, he says, "the yavanas call *sha.'s ifkari*", (6) Samrāṭ Yantra (Supreme instrument—an equinoctial dial), 'the best among the instruments' and (7) Jaya Prakāś 'the crest jewel of all instruments.'

Then we are told (in the verse quoted in full above) that Jagannāth prepared this excellent *Siddhānta Samrāj* for the delight of Jai Singh and that it is a rendering into Sanskrit for the benefit of mathematicians of a work in the Arabic language entitled *Mijāsti*.[1] He also tells us[2] that "in the Yavana country, the Yavana masters of astronomy, Abarkhas, etc., found the maximum declination to be 23 degrees 51 minutes 19 seconds; and that in Yunan, 36 degrees north, it was found to be 23 degrees 51 minutes 15 seconds by the observations of Vitlamayus. Ulugh Beg found it to be 23 degrees 30 minutes 17 seconds at Samarqand, 39 degrees 17 minutes north. By observation with this instrument we found it to be 23 degrees at Indraprastha in 1651 Sālivāhana."

(*ii*) Ulugh Beg's astronomical tables were completed in A. D. 1436 and became almost as famous as those of Ptolemy and they formed the basis of most subsequent catalogues. Flamsteed used them and so did Jai Singh, who brought them up to date. For details as to the use made by Jai Singh of Ulugh Beg's tables see below (p. 8).

(*iii*) The Hindu name for the astrolabe is *Yantra Rāja* and Garret says[3] that this "appears to be a very ancient type of instrument of Hindu origin", and also that "it appears to have been held in great esteem by Jai Singh as he himself wrote a book concerning its construction and use." As a matter of fact, the astrolabe or *Yantra Rāja* is not of Hindu origin at all. The earliest Hindu work on this instrument is of the fourteenth century of our era while numerous Arabic and Persian works dating from the tenth century onwards are known. The earliest Hindu work known is by Mahendra Sūri and was written in the time of Firoz Shah in Śaka 1292 or A.D. 1370[4] and there are indications that it was used by Jai Singh. But Jai Singh did not rely on this work alone

[1] *i.e.,* Ptolemy's *Almagest.*

[2] From the Calcutta MS. The names are somewhat puzzling but Abarkhas is for Hipparchus, Vitlamayus is for Ptolemæus. By Yunan possibly Rhodes is meant. The date 1651 Sālivāhana is equivalent to A. D. 1729. Indraprastha is Delhi.

[3] p. 49. See R. Mitra *Cat. Sans. MSS., Bikaner* p. 351.

[4] This work together with Malayendu's commentary was printed by the late Pandit Sudhākar Dvivedi of Benares. In the India Office Library is a manuscript (2935, 1528 a) of this work which was described by Eggeling (*Catalogue Sanskrit Manuscripts, India Office Library,* V. p. 1030. See also No. 2933 (2343 *b*) p. 1031) as follows :—

"*Yantra* rāja or *Yantrarājāgama*, also called *Suyantrāgama* and *Sadyantra*, a treatise in five chapters on the construction and use of the armillary sphere and the calculation of celestial and terrestrial longitudes and latitudes, by *Sūri Mahendra* the pupil of *Madana Sūri* the court astrologer of *Bhṛigupura.*'

and certainly used some of the Arabic or Persian text-books on the astrolabes, of which there were a great many available.

Mahendra describes his treatise as "This scientific work, the good *Yantra Rāja*, full of much variety and wonder causing, for the benefit of the people, etc." It is "abridged, essence-like, exhaustive but very simple and delightful to the heart." He says (v. 3): "Many *Yavanas* have also composed scientific works on this instrument in their own language and according to their own particular understanding" and, he continues, "having found them like oceans, I now compose this work, like nectar, as the essence of them all." He gives a list of thirty-two stars[1] and then (v. 28) writes: "After freeing these stars of drik karma mark them on the celestial globe . . . . This is a secret that has come from the Yavanas."

Mahendra's small star catalogue is of considerable interest because such lists are very uncommon in Hindu books and because it is taken from Ptolemy's catalogue. The latitudes are exactly the same as Ptolemy's in all cases but one, and the longitudes differ by exactly 18° 53′ in all cases but six.

(*iv*) Jai Singh himself refers to La Hire's tables (see page 14) and to other European tables, and in the palace library at Jaipur is still a copy of Flamsteed's great work.

P. de la Hire was a French scholar of repute who lived from A.D. 1640 to 1718. He wrote many mathematical works and in 1702 published his *Tabulae Astronomicae*, of which the first part had appeared in 1687. This work contained, besides the usual tables, a refraction table (which, it is said, Jai Singh copied) and a description of a machine invented by la Hire to show the theory of eclipses. Another of la Hire's works was 'La Gnomonique ou l'art de tracer des cadrans ou horologes solaires sur toutes sortes de surfaces, par differentes pratiques, avec les démonstrations géometriques de toutes les operations.' This was published in 1682 and would have been useful to Jai Singh.

(*v*) John Flamsteed lived from 1646 to 1720. His *Historia Cœlestis Britannica* appeared in 1712, in one folio volume, made up of two books, the first containing the catalogue of stars and sextant observations; the second, observations with Sharp's mural arc. The complete work, consisting of three folio volumes, was published in 1725. Flamsteed himself lived only long enough to finish the second of the three volumes. The third was edited by his assistants Crossthwaite and Sharp. It contains descriptions of the instruments used by Tycho Brahe, Hevelius, Flamsteed himself, etc.; the star catalogues of Ptolemy, Ulugh Beg, Tycho Brahe, the Landgrave of Hesse and Hevelius, and finally the British catalogue of 2,935 stars.

(*vi*) Undoubtedly Jai Singh possessed other astronomical works, but the only possible hints as to their identity are contained in the preface to his own catalogue where he mentions several astronomers by name. For example, he not only mentions Naṣīr-al-Dīn al-Ṭūsī (born A.D. 1201) but also his commentator (Ali b. M.) al-Gurgānī. Naṣir al-Dīn was one of the greatest Muslim astronomers.

[1] Mahendra's list is given in Appendix A. He gives the rate of precession as 54 seconds and it may be noted that $\frac{18° \ 53'}{54 \text{ secs. p.a.}}$ gives almost exactly 1250 years and this gives the date for Ptolemy's catalogue as A.D. 111 approximately.

He made observations at the Marāgha observatory and published the famous 'Ilkhāṇic Tables.' He wrote numerous works on astronomy and mathematics, including commentaries on the works of Archimedes, Euclid, Ptolemy, etc.

Coupled with Naṣīr al-Dīn, Jai Singh mentions also Jamshid Kāshi (Jamshid b. Mas'ūd b. M. Gijāt al-Dīn al-Kāshī) who was one of Ulugh Beg's assistants. He wrote several works on astronomy and particularly on the Kāqānī tables. He also mentions al-Ṣūfī (see page 10).

(*vii*) Hunter[1] tells us that he met at Ujjain a grandson of Jai Singh's principal assistant (? Jagannāth). "In his possession," he writes, "I saw the translation into Sanskrit of several European works, executed under the orders of Jaisingha, particularly Euclid's *Elements*[2] with a treatise on plane and spherical trigonometry, and on the construction and use of logarithms which was attached to Cum's and Commandine's edition. In this translation the inventor is called Don Juan Napier[3] . . . . Besides these the Pandit had a table of logarithms and of logarithmic sines and tangents to seven places of figures, and a treatise on conic sections."

We are also told that "maps and globes of the Feringhees were obtained from Surat."[4]

Personal assistance rendered to Jai Singh.

3. Jai Singh did not rely altogether upon information contained in books. He sent to Europe "several skilful persons along with Padre Manuel"; Muhammad Sharīf[5] was sent to some place where "the southern pole was overhead"; and Muhammad Mahdī was sent to the "further islands."[6]

Confirmation of the expedition to Europe is found in the records of the Jesuit Missionaries in India. In 1728 or 1729 we are told[7] that Jai Singh sent Father Figueredo, a Portuguese Jesuit, to Portugal. Also the same records relate that on January 6th, 1734, two priests set out from Chandernagore to Jaipur,[8] at Jai Singh's request. The account[9] of the astronomical work done by these two priests at Jaipur and on their journey was written, according to M. D' Anville,[10] by Father Boudier, one of the priests who made the journey. Obser-

[1] *Some account of the Astronomical Labours of Jaya Sinha, Rajah of Ambhere, or Jayanagar.* By W. Hunter. Asiatic Researches, Vol. V., 1799, p. 209.

[2] This is the *Rekhagaṇita* referred to on p. 69.

[3] This seems to be the source of Tod's statement that Jai Singh caused "Don Juan Napier on the construction and use of logarithms to be translated into Sanskrit." (ii. 358).

[4] Garrett p. 20. In the Jaipur museum there is a terrestrial globe attributed to Jai Singh; and for the transference of Ulugh Beg's co-ordinates into declination and right ascension a large and fairly accurate celestial globe was used by Jai Singh's assistants (see p. 8).

[5] There is a treatise on the astrolabe (British Museum Adit. manuscripts No. 7489) by 'Abdu'l Rahim b. Muhammad Sharīf al-Sharīf. The date of the manuscript is A. H. 1165 (=A.D. 1751). See Morley p. 2.

[6] Garrett p. 20.

[7] *Lettres édifiantes et curieuses, écrites des Missions étrangères. Nouvelle Édition. Mémoires des Indes. Tome quinzième. Tolouse* 1810 pp. 269 f.

[8] A journey of over a thousand miles.

[9] *Observations géographiques faites en 1734 par des Pères Jésuites, pendant leur voyage de Chandernagor à Delhi et à Jeypour* p. 169.

[10] *Eclaircissemens géographiques sur la Carte de l'Inde.* Paris 1753. p. 46. Father Boudier's account was not published till later, but M. D'Anville obtained the manuscript from M. Desp.éménil.

vations were made at most of the important places through which they passed. The observatories at Delhi and Jaipur are mentioned but not those at Benares and Mathurā, at both of which places they made astronomical observations, and this means that the observatories at Benares and Mathurā were probably built after their visit, which took place in the early part of 1734.

At the two observatories visited the following results were obtained :—

	Longitude East of Paris.		Longitude East of Greenwich.	Latitude north.
Delhi	75° 0′	=	77° 20′ 13″	28° 37′
Jaipur	73° 50′	=	76° 10′ 13″	26° 56′

From observations of an eclipse of the sun made on December 1st, 1732, by the Jaipur Pandits, Father Boudier calculated the difference in time between Paris and Jaipur as 4 hours 55 minutes 34 seconds east of Paris (=76° 13′ 43″ E. of Greenwich) and Father Boudier himself, observing the emersion of a satellite of Jupiter, calculated the longitude as 4 hours 55 minutes east of Paris (76° 5′ 13″ E. of Greenwich). In the *Lettres édifiantes et curieuses* we read (p. 239): "Les observations des satellites de jupiter ont été faites par le Révérend Pére Chaubil (? at Pekin) avec une lunette de vingt pieds, et par les Péres Jésuites qui étaient en voyage avec une de dix-sept pieds."

This visit is of such importance as to warrant quotations from early works regarding it. We read in the *Lettres édifiantes* (p. 269 f.): "Le Raja d'Amber, Jassing-Savaë, dont les Gazettes d'Europe firent mention en 1728 ou 1729, au sujet d'un voyage en Portugal, que le Révérend Père Figueredo, Jésuite Portugais, fit par ses orders, mourat en 1743........... Ce Prince ayant demandé des pères Jésuites de Chandernagore, l'espérance de le rendre encore plus favorable aux Chrétiens, en favour de qui il avait dèjà commencé une Église dans sa nouvelle ville,[2] détermina leur Supérieur-Général dans les Indes à lui en envoyer deux, qui partirent de Chandernagor de le 6 Janvier de l'année 1734, et qui firent les observations géographiques qu'on va rapporter. C'est tout ce que leur a permis de faire en ce genre l'incommodité des voyages en ce pays-ci, sourtout lorsqu'il faut les faire par terre, et leur mauvais santé, tous les deux devant leur retour ayant pensé mourir de maladie, causée par les fatigues et les mauvaises eaux qu'on est obligé de boire en chemin."

In 1775 M. D'Anville wrote[3] "Cet habile Astronome (P. Boudier) se rendant aux sollicitations d'un puissant Raja, nommé Jassing-savaë, fort curieux d'astronomie et qui non content d'avoir fait construire un observatoire dans la ville de sa résidence à environ cinquante lieues de Delhi, en avoit élevé un

---

[1] The Paris observatory is 0h 9m 20·9 secs =2° 20′ 13·5″ E. of Greenwich. Jaipur observatory is 75° 49′ 18·5″ E. of Greenwich, while Delhi observatory is 77° 13′ 5″ E. The approximately correct latitudes are Delhi 28° 37′ 35″ N., Jaipur 26° 55′ 27″ N.

In the latter part of the seventeenth century the difficulty of chromatic aberration was partially overcome by the use of very long telescopes—often a hundred feet or more. This led to 'aerial Telescopes' without tubes of which la Hire in 1715 gave a description: possibly Father Boudier's was a small one of this type.

[2] Jaipur was built about A.D. 1728.

[3] *Antiquité Géographique de l'Inde.* p. 60.

autre avec magnificence dans un de ces faubourgs, & appelé Jassingpura, met 3 minutes 40 seconds de difference entire la hauteur rapportée au Palais du Mogol et cet observatoire, ce qui donne un intervaile d'environ 4000 toises." He died about A.D. 1792.

Tieffenthaler, a French Jesuit, who landed in India in 1743, the year in which Jai Singh died, writes :[1] "J'ai fait trois ou quatre excursions à Agra et Delhi, pour faire visite au R. P. André Strobel, que Jessing, Raja de Djepour, curieux d'astronomie avoit appelé d'allemagne avec un compagnon."

The only other European connected with Jai Singh that we have information about is a Don Pedro de Sylva, who, according to Hunter[2], was a physician and an astronomer and resided at Jaipur with Jai Singh. De Sylva, it appears, died about A. D. 1792.

---

[1] *Description historique et géographique de l'Inde.* Ed. by J. Burnouilli 1786. Preface p. 5.

[2] p. 210.

Chapter II.—JAI SINGH'S ASTRONOMICAL TABLES.

4. *The Zīj Muhammad Shāhī* is a set of astronomical tables prepared under the direction of Jai Singh and named after the Emperor, Muhammad Shāh[1]. Of this work, I found (A) an incomplete Devanāgarī manuscript at Jaipur and (B) a complete Persian manuscript at the British museum. At first B was supposed to be an original work while A was said to be, not the *Zīj Muhammad Shāhī* itself but *Ulugh Beg's* celebrated catalogue brought up to date by Jai Singh and his assistants.

5. (A) The Jaipur manuscript begins as follows :—

> "Homage to holy Ganesh. Catalogue of 48 constellations. From the time of Ulugh Beg's table A.H. 841 to the present date A.H. 1138[2] or 297 years the mean motion[3] is 4 degrees 8 minutes. In the *Zīj Muhammad Shāhī* the estimates of declination, etc., are taken from "the globe. Right ascension divided by six is apparent time."

Two pages of the Jaipur manuscripts are shown in plate I (Figures 1, 2) and I give below extracts[4] from the manuscripts together with a table of comparisons. The manuscript gives : (*a*) The numbers of the constellations and star numbers and these in all cases follow Ulugh Beg's order exactly. (*b*) The nomenclature which is a translation from Ptolemy (through Ulugh Beg.) In a few cases the Persian and Hindu names are also given. (*c*) Ulugh Beg's longitudes with 4°8′ added for precession. (*d*) The latitude which in practically all cases is the same as Ulugh Beg's. (*e*) The so-called polar longitudes[5] (what Delambre calls 'false longitudes') this is the *Sūrya Siddhānta* method of indicating the positions of stars, but it also occurs in Muslim works, *e.g.*, Abū Ali al-Hasan in the 13th century of our era calculates the polar longitudes for a number of stars; and the presence of these polar longitudes in Jai Singh's catalogue is possibly due to Muslim and not Hindu influence. (*f*) Declinations and (*g*) Right ascensions apparently read off from a globe (*h*) Right ascension in *ghaṭis* and *palas* obtained from (*g*) by dividing by six.[6] (*h*) Star magnitudes which seldom differ from these recorded by Ulugh Beg.

The catalogue is not an original one but is Ulugh Beg's brought up to date.

---

[1] Hunter and others say that Jai Singh was chosen by Muhammad Shāh to reform the calendar but probably Jai Singh was the mover and at the most obtained the Emperor's formal sanction.

[2] A.H. 841 = A.D. 1437-8 ; A.H. 1138 = A.D. 1725-6 ; and 297 Muslim year = 288·2 Christian years. [A.H. 622—3 = A.D.]

[3] The precession of the equinoxes is meant. The rate here given, 4°8′ in 297 Muslim years, is equivalent to 51·6″ a year. See appendix D[7].

[4] Further extracts are given in appendix A.

[5] The polar longitude is marked on the ecliptic by the circles of declination, that is, the difference ($\Delta\lambda$) between the true longitude (λ) and polar longitude (λ') is that portion of the ecliptic intercepted between the star's declination and latitude circles. The polar latitude β' (which is not given in the Ms.) is, similarly, the part of the declination circle between the star and the ecliptic. The change of co-ordinates can be made by help of the following formulæ (*i*) $\tan M = \frac{\cot \omega}{\cos}$ (*ii*) $\sin \lambda' = \sin M . \sin \lambda'$ (*iii*) $\sin \Delta\lambda = \tan \lambda \cot M$.

[6] 60 palas = 1 ghati = 24 minutes = 6 degrees.

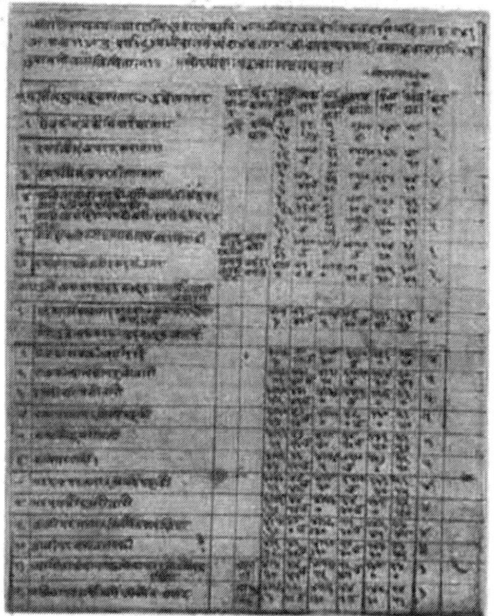

Fig. 1.

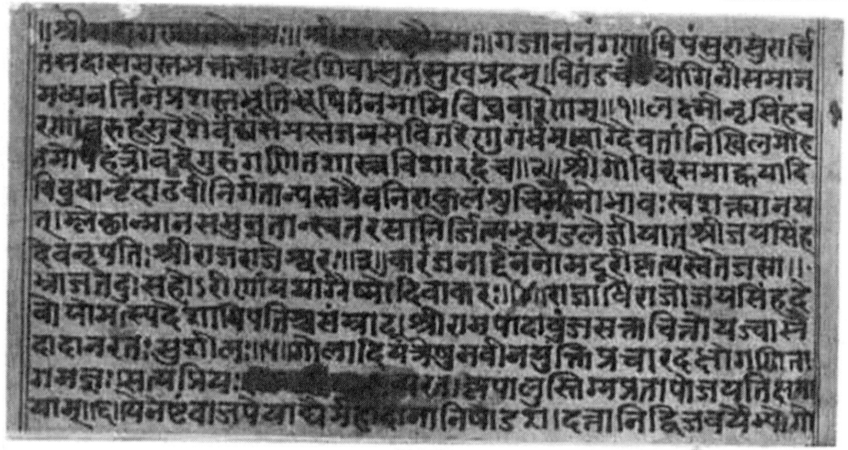

Fig. 2.

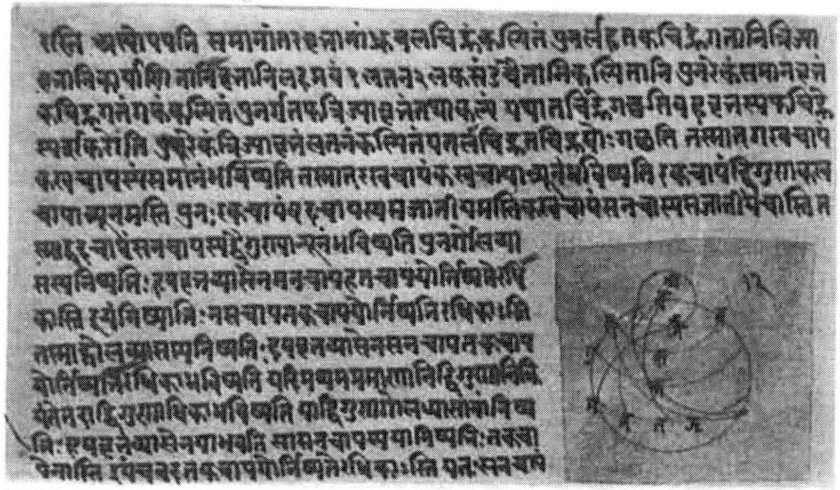

Fig. 3

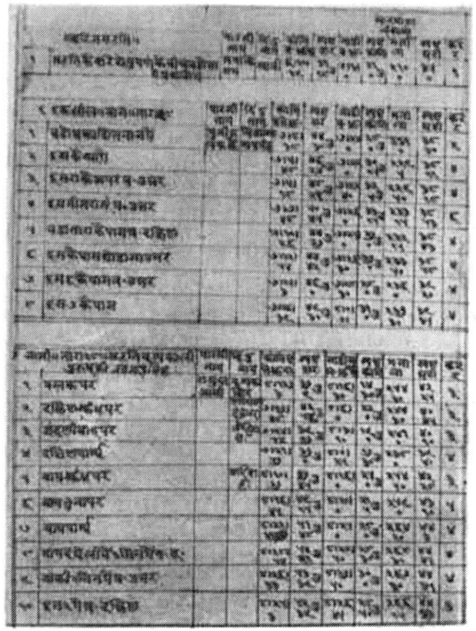

Fig. 4

Figs. 1 & 4 FROM MANUSCRIPT OF STAR CATALOGUE.
Figs. 2 & 3 FROM MANUSCRIPT OF SAMRAT SIDDHANTA

The method of transference of cordinates employed was bound to lead to errors of certain types, *viz.*, errors due to the graduations of the globe employed, and a greater apparent error in northern latitudes. The table has a special interest of its own and an interest in connection with Jai Singh's work as showing at least one of the sources of his astronomical knowledge.

EXTRACT FROM THE JAIPUR CATALOGUE.

A	B	C	D	E	F	G	H	I
		Longitude.	Latitude.	Polar Longitude.	Declination.	Right Degrees.	Ascension in ghatis and palas.	Magnitude.
VI	*Constellation of the Crown.*	s. ° ′	° ′	s. ° ′	° ′	° ′	g. p.	
1	Very brilliant .	7 8 38	+40 30	7 24 0	+28 0	231 0	38 30	2
2	Beyond this .	7 5 48	+46 24	7 22 0	+30 15	229 5	38 11	4
3	Above the second to the north.	7 5 18	+48 21	7 22 40	+32 5	229 50	38 18	4
4	The third to the north of this.	7 7 48	+50 45	7 26 0	+33 15	233 15	38 53	6
5	Near to the large star to the south.	7 10 26	+44 27	7 25 0	+27 0	232 15	38 43	4
6	Near this a little to the north.	7 12 54	+44 42	7 26 30	+27 0	233 50	38 58	4
7	Near to this the sixth to the south.	7 15 3	+46 0	7 28 30	+28 0	236 0	39 20	4
8	Near to number 7	7 14 39	+49 30	8 0 0	31 0	237 30	39 35	4

	From Ptolemy.	Modern name.	FROM ULUGH BEG.		DIFFERENCE BETWEEN MS. AND ULUGH BEG.		DIFFERENCE BETWEEN MS. AND FLAMSTEED REDUCED TO A.D. 1725.	
		1	Longitude.	Latitude.	Δ long.	Δ lat.	Δ long.	Δ lat.
VI	*Corona Borealis.*		s. ° ′					
1	Fulgens earum quæ sunt in corona. .	5 α	7 4 34	+44° 30′	+4°4	+0	+18′ 6″	−3°51′18″
2	Quæ omnes istas præcedit . . .	3 β	7 1 40	+46 24	+4°8′	0	+31′57″	+19′19″
3	Borealior quæ istam sequitur. . .	4 δ	7 1 10	+48 21	+4°8′	0	−17′30″	−13′51″
4	Sequens istam et borealior ista .	9 π	7 3 40	+50 45	+4°8′	0	−32′51″	+14′56″
5	Quæ fulgentem a meridie sequitur. .	8 γ	7 6 28	+44 27	+3°58′	0	−35′51″	−5′19″
6	Quæ istam proprius sequitur. .	10 τ	7 8 46	+44 42	+4°8′	0	−14′24″	−11′6″
7	Quæ post istas rursus sequitur. .	13 ε	7 10 55	+46 0	+4°8′	0	−13′25″	−6′28″
8	Sequens cunctas quæ in corona sunt.	14 ι	7 11 31	+49 30	+38′	0	−30′48″	+18′38″

6 (B). The British Museum manuscript[1] bears the title *Zīj Jadīd Muhammad Shāhī* (the new Muhammad Shāh tables) and Rāja Jai Singh Sawāī is indicated as the author. The work consists of three books: (i.) On the current eras, namely, the Hijrah, the eras of Muhammad Shāh, the Christian era, and the Samvat era. (ii.) On the determination of the ascendants. (iii.) On the motions of the planets and stars and their positions.

The first two sections follow Ulugh Beg and the third section is simply Ulugh Beg brought up to date. The catalogue of stars is headed: "Table showing the positions of the fixed stars as determined at the Samarqand observatory." The catalogue gives (*a*) serial numbers, (*b*) constellation numbers, (*c*) names of constellations and stars, (*d*) Longitudes, (*e*) Latitudes, (*f*) Directions, (*g*) Magnitudes according to Ptolemy, (*h*) Magnitudes according to Sūfī.[2] The total number of stars given is 1018 (Ulugh Beg's number) and these are arranged in identically the same order as those of Ulugh Beg. The latitudes are the same as Ulugh Beg's and the longitudes differ by 4° 8′, as in the case of the Jaipur MS.

7. The preface to the *Zīj Muhammad Shāhī* is, from historical point of view, perhaps the most interesting part and is here given in full.[3]

> Praise be to God, such that the minutely discerning genius of the profoundest geometers in uttering the smallest particle of it, may open the mouth in confession of inability; and such adoration, that the study and accuracy of astronomers who measure the heavens, on the first step towards expressing it may acknowledge their astonishment and utter insufficiency Let us devote ourselves at the altar of the King of Kings—hallowed be his name—in the book of the register of whose power the lofty orbs of heaven are only a few leaves; and the stars and that heavenly courser the sun, a small piece of money in the treasury of the empire of the Most High.
>
> If he had not adorned the pages of the table of the climates of the earth with the lines of rivers, and the characters of grasses and trees, no calculator could have constructed the almanac of the various kinds of seeds and of fruit which it contains. And if he had not enlightened the dark path of the elements with the torches of the fixed stars, the planets and the resplendent sun and moon, how could it have been possible to arrive at the end of our wishes, or to escape from the labyrinth and the precipices of ignorance.
>
> From inability to comprehend the all encompassing beneficence of His power, HIPPARCHUS is an ignorant clown, who wrings the hands of vexation; and in the contemplation of His exalted majesty,

---

[1] Rieu's *Catalogue of Oriental MSS.* "Add. 14373. Foll. 222; 11½ inches by 7¾; 12 lines, 4¾ inches long; written in Nastālik, with 'Unvan and gold ruled margins, apparently in the 18th Century (Francis Gladwin)." The MS. is in good condition and could easily be reproduced by rotograph.

[2] Abdarraḥmān b. 'Omar, Abū'l-Hosein, al-Sūfī (died A. D. 986) wrote on the fixed stars, the astrolabe, &c (H. Suter. *Die Math. u. Astr. d. Araber* p. 62).

[3] Hunter *As. Res. V.* p. 178 f.

PTOLEMY is a bat, who can never arrive at the sun of truth: the demonstrations of EUCLID are an imperfect sketch of the forms of his contrivance; and thousands of JAMSHĪD KĀSHĪ[1], or NAṢĪR ṬŪSĪ,[2] in this attempt would labour in vain.

But since the well-wisher of the works of creation and the admiring spectator of the theatre of infinite wisdom and providence SAWĀĪ JAI SINGH[3] from the first dawning of reason in his mind and during its progress towards maturity, was entirely devoted to the study of mathematical science, and the bent of his mind was constantly directed to the solution of its most difficult problems: by the aid of the supreme artificer he obtained a thorough knowledge of its principle and rules.

He found that the calculation of the places of the stars as obtained from the tales in common use such as the new tables of S'AID GURGĀNĪ[4] and KHĀGANI, and the *Tasahīlāt-Mula Chānd*[5] *Akbar Shāhī*[6], and the Hindu books, and the European tables,[7] in very many cases give them widely different from those determined by observation: especially in the appearance of the new moons, the computation does not agree with observation."

Seeing that very important affairs both regarding religion and the administration of empire depend upon these; and that in the time of the rising and setting of the planets, and the seasons of eclipses and of the sun and moon, many considerable disagreements of a similar nature were found—he represented it to his majesty of dignity and power, the sun of the firmanent of felicity and dominion, the splendour of the forehead of imperial magnificence, the unrivalled pearl of the sea of sovereignty, the incomparably brightest star of the heaven of empire, whose standard is the sun whose retinue the moon, whose lance is Mars and his pen like Mercury with attendants like Venus, whose threshold is the sky,

---

[1] Jamshid b. Mes'ud Gijāt al-Din al-Kāshi was one of Ulugh Beg's astronomers.

[2] Naṣir al-Din al-Ṭūsi was born A.D. 1201. He worked at the Marāgha observatory and published the famous 'Ilkhānic Tables.' He translated Euclid's *Elements* and Ptolemy's *Almagest*, and wrote many works on astronomy.

[3] Jai Singh writes in the third person.

[4] Possibly 'Ali b. M. al-Saijid al-Sarif al-Gurgānī, who lived from A.D. 1339 to 1414 in Shirāz, and wrote a commentary on Naṣir al-Ṭūsi's *Tadkira* (See H. Suter's *Die Mathematiker und Astronomen der Araber und Ihre Werke*, p. 172); but Gurgānī was a designation of Ulugh Beg's family and Ulugh Beg's tables were sometimes termed the Gurgānī canon [See L.P.E.A. Sédillot's *Prolégomènes des Tables astronomiques d'Oloug Beg*, p.c. xix; also *Āin-i-Akbari*, (iii) 20 and 41 (Jarrett's edition); *Akbarnāma*, (i), 204 (Beveridge's Edition).]

[5] Suter (p. 95) mentions one al-Khāqānī, an astronomer and astrologer who died in A.D. 1038 and who worked at improving the astronomical tables. The Khāqāni tables were supplementary to the Ilkhānic tables of Nasir al-Ṭūsī and were prepared and edited by Jamshid al-Kāshi.

[6] "Maulānā Chānd, the astrologer, who was possessed of great acuteness and thorough dexterity in the science of the astrolabe, in the scrutinising of astronomical tables, the construction of almanacs and the interpretation of the stars, was deputed to be in attendance at the portals of the cupola of chastity in order that he might observe the happy time and ascertain exactly the period of birth (of Akbar). He reported in writing to the exalted camp that according to altitudes taken by the Greek Astrolabe and by calculations based on the Gurgānī tables, etc." (*Akbarnāma*, Vol. I, 69-70 Ed. Beveridge). He also cast the horoscope of Jahāngir in A.D. 1570 according to the Greek canon (*ib.* ii, 506-7. See also i, 56 and 374).

[7] He is possibly referring to La Hire's *Tabulæ Astronomicæ* and Flamsteed's *Historia Cœlestis Britannica*.

whose signet is Jupiter, whose sentinel Saturn—the Emperor descended from a long race of kings, an ALEXANDER in dignity, the shadow of God, the victorious king MUHAMMĀD SHAH[1]: May he ever be triumphant in battle.[2]

He was pleased to reply, since you, who are learned in the mysteries of science, have a perfect knowledge of this matter, having assembled the astronomers and geometricians of the faith of Islām, and the Brahmans and Pandits,[3] and the astronomers of Europe, and having prepared all the apparatus of an observatory, do you so labour for the ascertaining of the point in question, that the disagreement between the calculated times of those phenomena, and the times which they are observed to happen, may be rectified.

Although this was a mighty task, which during a long period of time none of the powerful Rajas had prosecuted; nor among the tribes of Islām, since the time of the martyr prince, whose sins are forgiven[3], MIRZA ULUGH BEG, to the present, which comprehends a period of more than three hundred years,[4] had any one of the kings possessed of power and dignity turned his attention to this object. Yet to accomplish the exalted command he had received, he bound the girdle of resolution about the loins of his soul and constructed here[5] several of the instruments of an observatory, such as had been erected at *Samarqand*[6], agreeable to the Mussalmān books[7] such as *Zat al-Halqa*[8] of brass, in diameter three *gaz*[9] of the measure now in use and *Zat al-Sha'batain*[10] and *Zat al-Zaqatain* and *Sads*[11] *Fakhri* and *Shāmalah.*[12]

But finding that brass instruments did not come up to the ideas which

---

[1] Muhammad Shāh reigned from 1719-1748.

[2] This must have been written before 1739 when Nadir Shāh sacked Delhi.

[3] These seem to be curious expressions for a Hindu to use.

[4] Ulugh Beg was assassinated in A.D. 1449 while the *Zij Muhammad Shāhi* is supposed to have been published in 1728, approximately 297 Muslim years after. See p. 8.

[5] At Delhi.

[6] We have very little information about the observatory at Samarqand. Greaves stated that the quadrant used by Ulugh Beg was as high as the summit of St. Sophia at Constantinople, or about 180 feet. The earlier Muslim astronomers had also devised huge instruments. In A.D. 995 Abul Wafa used a quadrant of radius 21 feet 8 inches; al-Khojendi used a sextant with radius 57 feet 9 inches. Naṣir al-Din set himself the task of perfecting instruments, etc. See Sēdillot's *Prolégomènes des Tables astronomiques d'Ouloug-Beg*, p. cxxix. See also page 81 for a fuller account of Muslim instruments.

[7] See below for a bibliography of books on the astrolabe. There were numerous Arabic and Persian works on astronomical instruments available.

[8] A ring instrument, armilla, sphæra armillaris—(Nallino ii. 329).

[9] 3 gaz = 9 feet ordinarily, but perhaps here a *gaz* = 1 *danda* = 6 feet approximately.

[10] An astrolabe with two rings or parts. It is the *triquetrum* or *regulæ parallacticæ*. Al-Battāni calls it the 'long alhidade' (Nallino i. 321). In Leiden is a MS. *De ratione qua ope instrumenti Zat al-Sho'batain*, etc., by the celebrated *al-Kindi*. Muhammad bin 'Isāc b. Abi 'Abbād, Abu'l Hasan also wrote on the same instrument (Suter pp. 25 and 48).

[11] This must be the same as the *Shashtāmsa Yantra*, which, according to Jagganāth "the Yavanas called *sudsu/kari*." See pages 3 and 39.

[12] The Jai Prakāś is called *shamlah* by Hunter.

Fig. 5. OBVERSE OF ASTROLABE (JAIPUR A)

Fig. 6. OBVERSE OF ASTROLABE (JAIPUR B).

Fig. 7. REVERSE OF ASTROLABE (JAIPUR A).

[illegible] REVERSE OF ASTROLABE (JAIPUR B)

Photogravure. Survey of India Offices, Calcutta, 1917

he had formed of accuracy, because of the smallness of their size,[*] the want of division into minutes, the shaking and wearing of their axes, the displacement of the centres of the circles, and the shifting of the planes of the instruments, he concluded that the reason why the determinations of the ancients, such as HIPPARCHUS and PTOLEMY, proved inaccurate, must have been of this kind.

Therefore he constructed in Dār al-Khalafat Shāh Jahānābād,[1] which is the seat of empire and prosperity, instruments of his own invention such as *Jai Prakas* and *Rām Yantra* and *Samrāt Yantra*,[2] the semi-diameter of which is of eighteen cubits and one minute on it is a barley corn and a half[3]—of stone and lime of perfect stability, with attention to the rules of geometry and adjustment to the meridian and to the latitude of the place, and with care in the measuring and fixing of them so that the inaccuracies from the shaking of the circles and the wearing of their axes and displacement of their centres and the inequality of the minutes might be corrected. Thus an accurate method of constructing an observatory was established and the difference which had existed between the computed and observed places of the fixed stars and planets by means of observing their mean motions and observations was removed.

And in order to confirm the truth of these observations he constructed instruments of the same kind in Sawāī Jaipur, Muthra and Benares and Ujjain[4]. When he compared these observatories, after allowing for the difference of longitude between the places where they stood, the observations agreed.[5]

Hence he determined to erect similar observatories in other large cities so that every person who is devoted to these studies, whenever he wished to ascertain the place of a star or the relative situation of one star to another, might by these instruments observe the phenomena.[6]

But seeing that in many cases it is necessary to determine past or future phenomena; and also that in the instant of their occur-

---

* *Cf.* Alberuni *Chronology of Ancient Nations* (p. 11) who writes: "It is impossible to fix the parts of the greatest circle by means of the parts of the smallest circle. I refer to the smallness of the instruments of observation in comparision with the vastness of the bodies which are to be observed. On this subject I have enlarged in my book called *Kitāb-al-istishhād bikhtilāf-al 'arsād.*" L.P.E.A. Sédillot [p. cxxix] gives the following interesting quotations: "Si j'avais pu, disait Ebn-Carfa, faire un cercle qui s'appuyat d'un coté sur les Pyramides et de l'autre sur le mont Mocattam, je l'aurais fait; car plus l'instrument est grand, plus les operations sont justes."

1 *i.e.*, Delhi.

2 See below p. 36 seq.

3 To make the measurements fit the cubit used must have been a large cubit – 36 aṅgulas.

4 This implies that the Delhi Observatory was completed before the others were started; and that all of them were built before the preface was written. This dates the preface after 1734 and perhaps after 1737. (See p. 15).

5 We must accept these statements about perfect agreement with some caution. We have very few records of Jai Singh's actual calculations or observations: his value for *precession* was 51·6″ a year and for the *obliquity* 23°28′0.

6 The project of building observatories at other places was never carried out.

rence cloud or rain may prevent the observation—or the power and opportunity of access to an observatory may be wanting—he deemed it necessary that a table be constructed by means of which the daily places of the stars being calculated every year and disposed in a calendar may always be in readiness.

In the same manner as the geometers and astronomers of antiquity bestowed many years on the practices of observation—thus, for the establishment of a certain method, after having constructed these instruments, the places of the stars were daily observed.

After seven years had been spent in this employment[1] information was received that about this time observatories had been constructed in Europe[2] and that the learned of that country were employed in the prosecution of this important work—that the business of the observatory was still carrying on there and that they were constantly labouring to determine with accuracy the subtleties of this science.

For this reason, having sent to that country several skilful persons along with PADRE MANUEL[3], and having procured the new tables which had been constructed there thirty years before and published under the name LIR[4] as well as the Europe tables anterior to those,[5]

On examining and comparing the calculations of these tables with actual observations it appeared that there was an error in the former in assigning the moon's place of half a degree. Although the error in the other planets was not so great, yet the times of solar and lunar eclipses he found to come out later or earlier than the truth by the fourth part of a *ghaṭi* or fifteen *palas*.[6]

Hence he concluded that since in Europe astronomical instruments have not been constructed of such a size and so large diameters, the motions which have been observed with them may have deviated a little from the truth.[7]

Since in this place by the aid of the unerring Artificer astronomical ins-

---

[1] The chronology is very uncertain. Delhi Observatory was constructed probably about 1724 and the tables, it is said, were finished in 1728; but there is evidence that leads us to the conclusion that this preface was written later. (See p. 139).

[2] Uraniborg (Tycho Brahe's observatory) in 1576; Leiden 1632; Paris 1667; Greenwich 1675; Berlin 1705; St. Petersburg 1725; Upsala 1730, etc.

[3] In 1728 or 1729 the Reverend Father Figueredo, a Portuguese Jesuit, went to Europe by the order of Jai Singh. Possibly this is the same man. See *Lettres edifiantes et curieuses*, xv. 269.

[4] La Hire's *Tabulae Astronomicae* was published in 1702; see p. 4 Father Boudier, who went to Delhi and Jaipur in 1734, actually refers to this edition. He writes: En se servant de la méthode de M. de la Hire, édition de ses tables 1702, page 53, on a trouvé que le commencement de l'éclipse à Delhi, lorsqu'il était à Rome 11 heures 40 minutes 55 secondes du matin, etc.," *Lettres*, etc. xv, 288.

[5] We know that, besides La Hire's tables, Jai Singh possessed those of Ulugh Beg and Flamsteed. The latter's work contains also the tables of Tycho Brahe, the Landgrave Hesse and Hevelius. Other possible tables are the *Toletan Table* of 1080; the *Alfonsine Tables*, 1252; Reinhold's *Prussian Tables*, 1551; Kepler's *Rudolphine Table*, 1627; Cassini's tables, 1668 and 1693; Halley's tables, 1719; etc.

[6] 60 *palas* = 1 *ghaṭi* = 24 minutes, and 15 *palas* = 6 minutes.

[7] The instruments used by Flamsteed (1646-1719) were an iron sextant of 6 feet radius; a three-foot quadrant; a mural arc of 140 degrees and radius 7 feet 'divided with hitherto unapproached accuracy' and with which all his most valuable work was executed; two clocks and two telescopes. For further particulars of European instruments, see p. 83.

truments have been constructed with all the exactness that the heart can desire and the motions of the stars have for a long period been constantly observed with them, agreeable to observations mean motions and equations were established; he found the calculation to agree perfectly with the observation. And although to this day the business of the observatory is carried on, a table under the name of His Majesty, the shadow of God, comprehending the most accurate rules, and most perfect methods of computation was constructed—so that when the places of the stars and the appearance of the new moons and the eclipses of the sun and moon and the conjunction of the heavenly bodies are computed by it they may arrive as near as possible to the truth, which, in fact, is every day seen and confirmed at the observatory.

It therefore behoveth those who excel in this art, in return for so great benefit, to offer up their prayers for the long continuance of the power and the prosperity of so good a King,[1] the safeguard of the earth, and thus obtain for themselves a blessing in both worlds[2].

---

[1] Muhammad Shāh died in A.D. 1748 five years after the death of Jai Singh.

[2] There are some points about the preface that are not quite consistent with each other and known facts. The tradition is that the *Zij Muhammad Shahi* was completed in A.D. 1728 and this is, to some extent, confirmed by the Jaipur MS.; the preface was written some time after all the observatories had been built, that is after 1734, and "more than 300 years" after (the death of) Ulugh Beg [Ulugh Beg died in A.H. 853, and 853+300=1153 A.H. = A.D. 1740-1]. The legitimate conclusion is that the preface was written some considerable time after the tables had been completed.

In 1902 Garrett wrote of the *Zij Muhammad Shahi*: "I have been unable to procure the Sanskrit original or even a vernacular copy" and "up to the present time the only copy of Jai Singh's astronomical tables, or *Zeech Mahommed Shahi*, which has been obtained is a book in Persian characters ...... Unfortunately most of the figures are written in a kind of cypher, and although the key to this has been found, the thorough examination of this work will necessarily prove a long and laborious task." (pp. 18n. and 74). The British Museum Persian MS. is in excellent condition and although the tables are, of course, in the *abjad* notation, there would be no difficulty in translating them.

Chapter III.—METAL INSTRUMENTS.

8. Jai Singh himself tells us that he first constructed, 'according to Mussulman books' instruments of brass such as *Zāt al-Ḥalaq*, *Zāt al-Sho'batain*, etc., and at Jaipur I found a unique collection of such instruments, including Arabic and Persian astrolabes, dating from the time of Shāh Jehān. These instruments play a very important part in Jai Singh's work; to appreciate which a proper understanding of them is essential. Enquiries in other parts of India resulted in the discovery of an excellent astrolabe in the Indian Museum, Calcutta; and one of rather inferior workmanship at Lahore. Tod tells us[1] of a dial "on the terrace of the palace of Oodipoor, and various instruments at Kotah[2] and Boondi, especially an armillary sphere, at the former, of about five feet in diameter, all in brass, got up under the scholars of Jey Singh." At the Lahore Exhibition of 1864 certain brass astronomical instruments from Kapūrthala and other places were shown: these included "two fine astrolabes," one spherical and one plane, and several dials.

The metal instruments actually examined, most of them at Jaipur, were as follows:—

A.—Astrolabe. Diameter 13 inches. Seven tablets. Jaipur. Figures 5 and 7.

B.—Astrolabe dated the 31st year of the reign of Shāh Jehān and A.H. 1067 (=A.D. 1657). Diameter 13 inches. Jaipur. Figures 6 and 8.

C.—Astrolabe. Designed by Muhammad Amīn bin Muhammad Ṭāhir and engraved by A'bdul Alimah. From Herāt. Diameter 7·3 inches. Indian Museum, Calcutta. Figures 9, 12, 15 and 16.

D.—Astrolabe. Diameter 6 inches. Jaipur. Figures 11 and 14.

E.—A Zarqālī astrolabe dated the 23rd year of the reign of Aurangzīb and A.H. 1091. (=A.D. 1680). Made for Nawāb Iftikhar Khān by a certain Zia-al-Dīn. Diameter 2 feet. Jaipur. Figures 19, 20, 21, 22.

F.—Astrolabe. Brass, 4·3 inches in diameter. Lahore Museum.

G.—Hindu Astrolabe. Diameter 16 inches. Jaipur. Figures 26 and 27.

H.—Jai Singh's iron *Yantra Rāj*. Diameter 7 feet. Figure 28.

I.—Jai Singh's brass *Yantra Rāj*. Diameter 7 feet. Figure 29.

J. *Unnatamśa Yantra*. A graduated brass circle 17½ feet in diameter. Jaipur.

K.—*Chakra Yantra*. There are two at Jaipur 6 feet in diameter and one at Benares 3 feet 7 inches in diameter. Figures 57 and 65.

L.—*Krānti vṛitti Yantra*. Jaipur. Figure 58.

M.—Hindu Astrolabe. Jaipur.

N.—*Dhruva-bhrama Yantra* or 'Circumpolar instrument' Jaipur.

---

[1] Vol. II, 359.

[2] See p. 34 for an account of an instrument presented by the Rajah of Kotah to the Government of India.

PLATE III.

Fig. 9. OBVERSE OF ASTROLABE (HERAT C).

Fig. 10. OBVERSE ASTROLABE (JAIPUR D).

Fig. 11. REVERSE OF ASTROLABE (JAIPUR D).

Fig. 12. REVERSE OF ASTROLABE (HERAT C).

Photogravure—Survey of India Offices, Calcutta 1917.

O.—Armillary sphere at Jaipur. One also at Kotah.

P.—Arabic Astrolabe. Brass, 5·7 inches in diameter. Delhi.

Q.—Persian Astrolabe. Brass gilt, 3·75 inches in diameter. Delhi.

R.—Hindu Astrolabe. Copper, 7 inches in diameter. Delhi.

Of these it is possible that all except 'C' belonged to Jai Singh and it is pretty certain that a number of his instruments have been lost (*e.g.*, see page 31). The most important are A and B (which, for convenience, I term 'Jaipur A' and 'Jaipur B') and the Zarqālī instrument E. 'Jaipur A' and 'Jaipur B' are of extremely fine workmanship, while E is an interesting example of a type hitherto seldom described in detail in European works.

The Astrolabe.

9. Of these metal instruments the astrolabe appears to have played the most important part in Jai Singh's work. Indeed in the middle ages the astrolabe was one of the chief astronomical instruments. The Arabs perfected it at a very early date and it remained one of the principal astronomical instruments until about the 17th century, and is still used in the East for astrological purposes. It was usually of brass[1] and varied in diameter from a couple of inches to several feet. The mariner's astrolabe (as used by Columbus) was adopted from that of the astronomers about A.D. 1480 but was superseded by Hadley's Quadrant of 1731. The famous scholar Gerbet, who afterwards became Pope Sylvester ii, had such skill in making astrolabes, etc., that he was supposed to have sold his soul to the devil. There are many references in mediæval literature to the astrolabe. More than three centuries before Jai Singh, Chaucer wrote his *Treatise on the Astrolabe*. "Trust well," he says, "that alle the conclusions that have be founde, or else possibly might be founde in so noble an instrument as is an Astrolabe been unknowe parfitly to any mortal man in this regioun, as I suppose."

10. The type of astrolabe principally used by Jai Singh was the flat astrolabe or *astrolabium planispherum*, in Arabic called *Zāt al-Ṣafā'iḥ* ('Consisting of tablets') like 'Jaipur A' and 'Jaipur B', to which the following description particularly applies.

The *corpus astrolabii* is a circular disc with a raised edge into which fit the several parts of the instrument:

(*i*) The containing disc is termed the *mater*[2] (*Ar. umm*) and the inner part of this is the venter[3] while the raised edge is called the *kuffa* or rim.[4] The venter is often inscribed with latitudes and longitudes of important cities (Figures 13, 14, 15).

---

[1] Gower refers to one of gold: "With him his astrolabe he name, which was of fine gold precious, with points and circles marveillous."

The Granada astrolabe described by H. S. Cooper (*JRAS* 1904, 53f) has silver knobs on each pointer of the '*ankabūt*'; in the British Museum are several inlaid with silver; and others evidently had some sort of jewel fixed in the *kursi* (see fig. 12) Gilt instruments are not uncommon.

[2] Moder, mother, rotula.

[3] Also *wajh* or face.

[4] Also called Margilabrum or Limbus, *Ḥajra* (side) etc.

(*ii*) The *'ankabūt* or *aranea*[1] is an open work disc marked with the ecliptic the signs of the zodiac and a number of stars. It is placed in the venter and can be revolved. The branches on which the names of the stars are written and the points of which indicate the positions of the stars are termed *shaẓāīā* or 'splinters.'[2] The pointer at the top of the 'ankabūt at the first point of Capricorn is termed the *muri* or index.[3] (Figures 5, 6, 9 and 10).

(*iii*) Several thin discs or tablets[4], marked with *almucantarats*[5], azimuth circles, hour circles, etc., for various latitudes, etc., fit into the body of the astrolabe. (Figures 17 and 18).

(*iv*) The alhidade or[6] sighter revolves round the centre on the back of the *mater*. Each arm has a perforated *libna*[7] or 'tile' which is sometimes hinged on to the alhidade. European astrolabes sometimes had another marker or label[8] without sights for use on the front of the instrument. (Figure 7.)

(*v*) The tablets and alhidade, etc., are fixed together by a pin (Ar. *quṭb*[9] which is fastened by a wedge termed by the Arabs *faras* or 'horse,' and often fashioned into some resemblance of a horse's head.[10] (Figure 24).

(*vi*) The whole is suspended by a ring (Ar. *halqa*) joined to the *'urwah* or handle, which in its turn is riveted to the projecting part, *kursi* or throne, of the mater. To the halqa was sometimes attached a cord (Ar. *'ilāqa*).

(*vii*) The back of the astrolabe (*Ẓahr al-usṭurlāb*) in all cases has an outer graduated scale, two upper quadrants and certain shadow scales. It is often inscribed with tables of use to the astrologer and geographer: the details vary greatly. (Figures 7, 8, 11 and 12).

The sighter and graduated circle (fig. 7) on the back of the astrolabe form the part of the instrument used in actual observation; while the 'tablets' and the '*ankabūt* (which rotates) and the graduated circle on the raised edge (*kuffa*) of the *mater* form a very efficient calculating machine.

The Tablets (Safaih).

11. The ordinary disc or 'tablet' is marked on each side with stereographical projections of the horizon and almucantarats, azimuth and hour circles for a particular latitude and also the equator and tropics.

---

[1] Also *shabakah*, net, rete; Alancabuth; Volvellum, etc.

[2] *Ibr al-Kawākib*, needles of the stars, etc.

[3] *Muri rās al-jadī* (index of the head of Capricorn); Almury, Ostensor, Denticle, etc. "Thin Almury is cleped the Denticle of Capricorne. This same almury sit fix in the hed of capricorne." Chaucer, I, 23.

[4] *Ar. Safā'iḥ*, Saphiahs, Tympana, Tabulæ regionum, etc.

[5] *Circuli progressionum* (Ar. *muqanṭarāt*).

[6] *Ar. 'iḍāda*, 'door post'; Dioptra, Mediclinum, Verticulum, Alidade, etc. Tanner (1587) describes the alhidade thus: "Altriada or mediclinium, in which are put two little pins or tables to take the height of the sun in the day and of the stars at night, of which one side goeth through the centre of the astrolabe is called the line of trust, because it bringeth credit of things practised there."

[7] Tabella, Pinna, Pinnula; *hadaf*, *dafa*.

[8] Ostensor, index, petite roue, etc.

[9] Axis, clavus, exiltre, alchitot, cavilla, etc; *miḥwar* (axis), *watad* (stake).

[10] See D and E (fig. 24). The former (D) is from the India Office Persian astrolabe and E is taken from the British Museum MS. of Mās'ā Allah's work. See also page 63 and figure 68; and Nallino al-Battānī I, 319.

Fig. 13. VENTER OF ASTROLABE (JAIPUR B)

Fig. 14. VENTER OF ASTROLABE (JAIPUR D)

Fig. 15. VENTER OF ASTROLABE (HERAT C)

Fig. 16. TABLET OF HORIZONS (HERAT C)

Photogravure. Survey of India Offices Calcutta, 1917.

The **almucantarats** are circles of altitude,[1] the almucantarat of zero altitude being the horizon (*EHW* in plate V). In the diagram almucantarats are drawn for 60°, 40°, 20° (as well as the horizon, 0°) and are marked a_{60}, a_{40}, a_{20}. The number of almucantarats varies in different astrolabes: if there is one for each degree of altitude the instrument is called *tamm* or 'complete'; if for every other degree it is termed *nisfi*, 'bipartite' and so on. Jaipur A and B (figures 5 and 6) are *tamm* or 'complete'; Herāt C (figures 17 and 18) is *nisfi*; while 'Jaipur D' (figure 10) is *sudsi* or sexpartite.

Azimuth lines are drawn at right angles to the almucantarats.[2] These are seen in figures 17 and 18 and in plate V portions of certain azimuth circles Z_{58}, Z_{42}, etc., are drawn. In some tablets these azimuths are continued below (to the north of) the horizon.

Temporal hours.—The temporal or unequal or planetary hour lines are shown in figures 17 and 18 and in plate V by the broken lines t_1 t_1, t_2 t_2, etc.[3] They divide the time between sun-rise and sunset into twelve equal portions and therefore vary in length from day to day. These divisions of time gradually fell into disuse (see page 87) and equal or **equinoctial hours** were introduced. These are shown in figures 18 (but not in figure 17) and in plate V they are marked e_1 e_1, e_2 e_2, etc.

Houses.—The tablet is sometimes divided into twelve astrological 'houses.' The boundary lines of these are seen in figure 17 and in plate V are marked h_1 Hh_7, h_2 Hh_8, etc.[4] (See appendix B).

Longest days and latitudes.—The latitude for the particular tablet is gene-

---

[1] Let *ABCD* (Plate V) represent the tropic of Capricorn and *CA* the meridian, and let the arc *AF* measure the obliquity of the ecliptic, then the point S on the intersection of *BF* and *AC* is on the equator and *SENW* represents the equator, *E* being the eastern point. Similarly by drawing *Os* parallel to *OF* we get *s*, the southern point of the tropic of Cancer and *sC* the diameter of the ecliptic. The angles *SL* and WL_1 measure the latitude (ϕ) of the place, and by joining *LE* and L_1E we get Z, the zenith, and *H*, the meridian point on the horizon. The opposite point on the horizon is H_1 where EL_2 (=ϕ) meets the meridian line *NS* produced. To obtain a circle (almucantarat) for altitude *a* mark off angles (ϕ±*a* 90°) from *S*, the south point of the equator, (positive direction *SW*) and join both these points to *E*, the east point on the equator: the distance between the points intercepted on the meridian line *NS* is the diameter of the circle of altitude *a*.

[2] In the diagram $NL_2 = SL = \phi$, the latitude of the place, and EL_2 cuts the meridian line *NS* in the nadir *n*. The horizon is graduated by joining the zenith, *Z*, and the graduations on the equator and each azimuth circle passes through the zenith, nadir and the point on the horizon to which it pertains, while the centres of the azimuth circles lie on the line parallel to *EW* and bisecting *Zn*.

[3] To draw the temporary hour circles divide the day portion (that is the portion below the horizon in the diagram) of each of the three circles—the tropic of Capricorn, the equator, and the tropic of Cancer—into twelve equal parts and draw circles t_1 t_1, t_2 t_2 etc., through each trio of corresponding parts. For the equal hours draw through *X*, the centre of the circle of the horizon, a circle concentric with the equator and tropics. Graduate this circle at intervals of 15 degrees starting from the south point and proceeding westwards. With the south point and in succession the other points of graduation as centres, and radius equal to the radius (*XH*) of the horizon draw arcs e_1e_1, e_2e_2, etc., from the circle of Capricorn to that of Cancer. (These arcs will pass through the equal divisions of the equator, already marked for temporal hours). The result of this construction is that any point on the ecliptic, as the 'ankabūt is rotated, passes from one equal hour line to another in one twenty-fourth of a revolution.

[4] The lines that divide the houses pass through *H* and H_1, the points common to the horizon and meridian, and points on the Equator at intervals of 30 degrees starting from the East and West line. Their centres lie on the line that passes through X (the centre of the circle of the horizon) and is parallel to the East and West line. The points of intersection of these house lines with the Ecliptic are termed cusps. (For further details see p. 120 and STOFLER *Elucidatio fabricæ ususque astrolabii* 1524.)

rally written just below the *ufk* or oblique horizon, *EWH*, on the right of the meridian line; while in the corresponding place on the left of the meridian, the length of the longest day of the year for the particular latitude is generally given. In figure 18 for example we have

Hours	Latitude
13 47	28

In some cases the name of a city is also given. For example 'Delhi R' has "Avaṁtikayām 22" and "Amadāvād' 23."

(*vii*) *Special Tablets.*—In the ordinary astrolabe the number of tablets varies to as many as nine—not counting the 'ankabūt. Generally one is a special disc for horizons on one side and celestial co-ordinates on the other and occasionally there are other special tablets: the rest are the ordinary tablets, already described for several latitudes.

Horizons.—The tablet of the horizons (*al-Ṣafīḥah al-āfāqīyah*) is shown in figure 16. The horizons are arranged in four sets—one set in each quadrant consisting of six or seven horizons—and below each of these sets are two scales termed *al-mail al-kulli* (*shamāli* or *janūbi*) or the total obliquity (northern or southern).

'Ankabut co-ordinates.—On the other side of the 'tablet of the horizons' is generally the 'tablet of the latitude of the complement of the total obliquity.' This really gives the celestial co-ordinates (longitude and latitude) and by its aid the positions of the stars on the 'ankabūt can be at once read off. Morley seems to have thought[2] that this was an ordinary tablet for latitude 66½° N. but it is quite rightly described as *Ṣafīḥah mizan al-'ankabūt*, 'the tablet of the measure of the 'ankabūt.' It occurs in most of the astrolabes I have examined.

(*ix*) The Lahore Museum astrolabe has a tablet for the latitude of the equator. Here the oblique and straight horizons coincide and this line is marked *al-maghrib al-arḍ lah* 'the west—no latitude' and *al-mashrik al-arḍ lah*, 'the east no latitude.' Below the horizon are concentric semicircles which appear to be circles of declination: a similar tablet is in the 'Delhi Q' instrument.[3] Another curious tablet belongs to the Jaipur Hindu *niṣfī* astrolabe (described below p. 31) and is for latitude 72° North. It has engraved upon it only the three circles and the almucantarats (none being numbered—the only number inscribed being '72.') The India Office Hindu astrolabe contains a similar projection, but more complete and is marked *aṁśa* 72, *horā* 23.

[1] Avaṁti is Ujjain, the latitude of which is roughly 23° 10′. The latitude of Ahmodabad is approximately 23° 6′.

[2] Morley says: "These last two mentioned *Ṣafīḥahs* appear to have been used as models for the construction of the ordinary tablets" but he is not altogether right in the second case.

[3] Morley (pp. 12—13) describes three tablets for 'no latitude' and gives diagrams (XX, 17, 18 and 19). One belongs to the India Office Persian Astrolabe, another the Vaux astrolabe, the third is the India Office Hindu astrolabe.

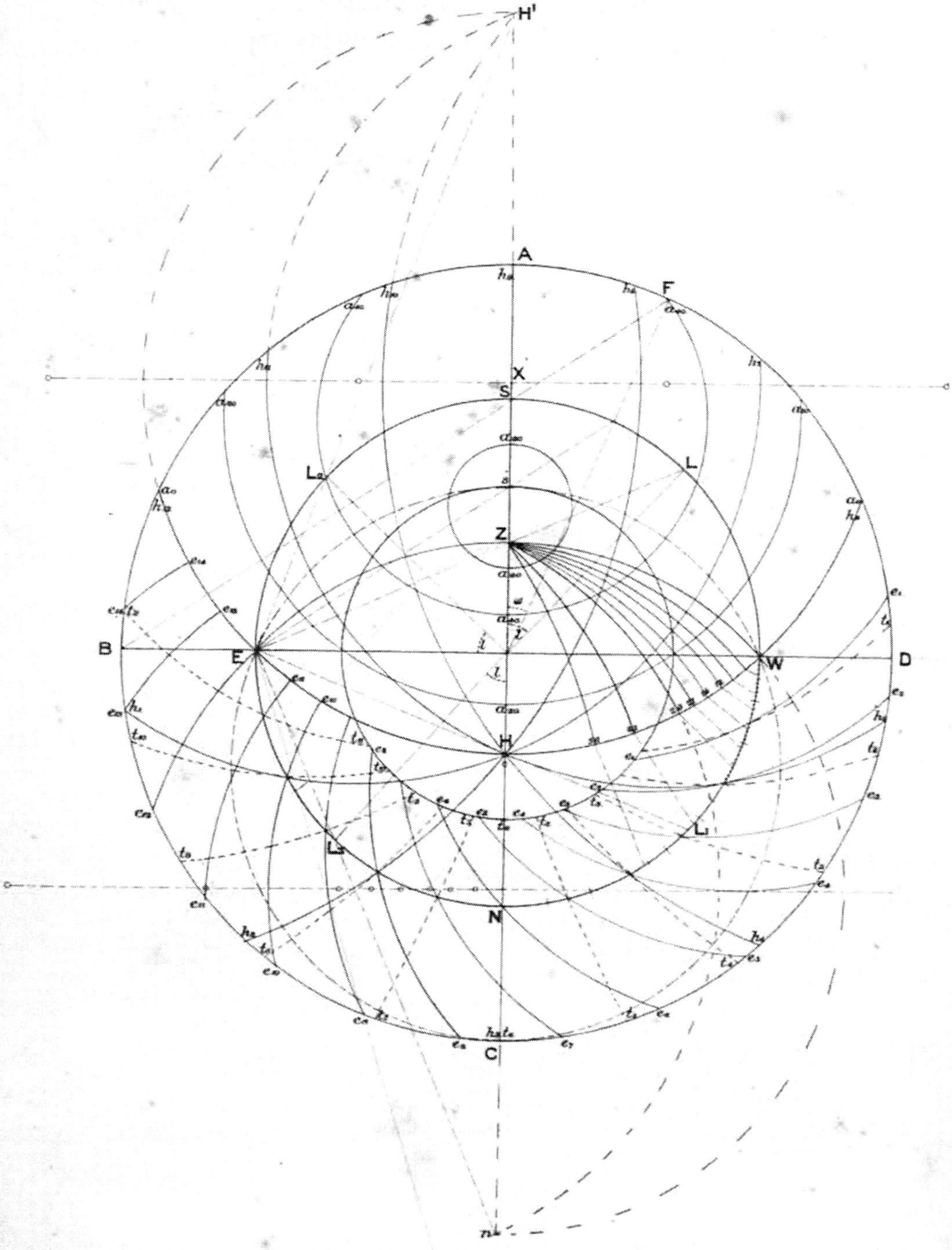

CONSTRUCTIONS FOR ASTROLABE TABLET.

In an astrolabe (Delhi Q) that was shown to me in Delhi there are two other types, (*i*) a tablet with co-ordinates for latitudes 30° ($13^h\ 56^m$) and 28° ($13^h\ 46^m$) engraved on the same side, and a similar one for latitudes 40 ($14^h\ 51^m$) and 62°, (*ii*) a tablet split into two halves along the meridian, for latitudes 32° ($14^h\ 6^m$) and 36° ($14^h\ 26^m$).

12. **'Ankabut** (*aranea*) or **shabakah** (*rete*). The '*ankabūt* or 'spider' is an open work tablet so arranged that the one below it may be conveniently seen (Figures 5, 6, 9 and 10). It exhibits a graduated ecliptic circle[1] with the signs of the zodiac and a number of the more important northerly stars. The points (*shaẓāīā*) of the net work indicate the positions of the stars, the names of which are engraved on the branches. The number of stars varies with the size, etc., of the instrument. The small Jaipur astrolabe (*D*) has 25, the Shāh Husain instrument has 63, etc.

The 'ankabūt is generally the most ornamental part of the instrument: it is sometimes inlaid with silver; Jaipur A shows the forms of the constellation animals, etc. The 'ankabūt is not used in a fixed position like the other tablets, but can be rotated and thus is employed, in combination with the tablet placed below, for finding the position of any star at a given time, the ascendant or 'horoscope,' the time and length of the day, etc., etc.

13. **Back of the Astrolable** (*Zahr al-Aṣṭurlāb*).

The back of the astrolabe is usually covered with a great deal of information, useful principally to the geographer and astrologer. The several instruments differ in detail, but the general arrangement is much the same. The contents may be roughly classified thus:—

(*a*) The upper half of the periphery is graduated into degrees, etc.

(*b*) The South East quadrant[2] consists of a graphic table of sines.

(*c*) The South West quadrant[2] is inscribed with declination graphs, etc.

(*d*) Shadow scales (lower periphery and central rectangle).

(*e*) Tables of signs, mansions, planets, terms, faces, etc.; generally contained in the inner semi-circles of the lower half of the disc.

Special tables contained in rectangles such as:

(*f*) The times of the rising of the signs (In the centre of figure 8).

(*g*) Trigons or triplicates and their regents, (In the centre of figure 12 and right of figure 7).

(*h*) Table of climates, (Lower part of figure 7).

(*i*) Differences between true and nominal years (Left of figure 7).

(*a*) The periphery of each of the upper quadrants is graduated into degrees commencing from the east and west points. In 'Jaipur A' (figure 7) the degrees are divided into quarters. In conjunction with the alhidade or sighter these graduations were used for measuring altitudes and other angles.

---

[1] The diameter of the ecliptic is the distance Cs or At_4 (Plate V) between the intersections on the meridian line N S of the tropics of Capricorn and Cancer. The graduations of the ecliptic lie on the line joining the pole of the ecliptic to the corresponding graduations on the equator. The pole of the equator is, of course, at the centre of the disc, while the pole of the ecliptic is at a distance from the pole of the equator equal to the maximum declination (approximately 23½°). It is sometimes marked *Quṭūb ul-buruj*.

[2] The south point is at the top of the disc.

(*b*) The 'quadrant of sines' occupies the south-east quadrant and occurs in most instruments. In some instruments the vertical radius is divided into sixty equal parts and lines parallel to the other radius are drawn to the circumference from each point of division (figure 12); in others both radii are so divided and horizontal and vertical lines are drawn from each point of division (figure 8); in others horizontal lines are drawn from each degree on the quadrant. The vertical and horizontal scales indicate the sines and cosines of the corresponding angles. In the description of the Zarqālī astrolabe below (p. 29) the use of these scales is explained in more detail. In some instruments arcs for 23½°, 30°, etc., are described.

(*c*) The south-west quadrant in most instruments exhibits a sort of yearly calendar. The horizontal and vertical radii are divided into six equal divisions[1] figures *7, 8*). From the points of division arcs are described and the names or numbers (figure *11*) of the signs are written in the spaces, six on the horizontal radius and six on the vertical, in the following order—

	3	4	5	6	7	8
Vertical . .	Cancer	Leo	Virgo	Libra	Scorpio	Sagittarius
Horizontal .	Gemini	Taurus	Aries	Pisces	Aquarius	Capricornus
	2	1	0	11	10	9

This division combined with the graduated circumference forms a scale of circular and angular co-ordinates and on this scale are traced various kinds of graphs[2] showing for example—

(*i*) The relation between the sun's right ascension and meridian altitude (figures 7, 8 and 37);

(*ii*) The meridian altitudes for certain latitudes (figure 12);

(*iii*) The altitude of the sun when it traverses the azimuthal circle of the Ka'bah at Mecca;

(*iv*) The temporal or unequal hours (figure 11).

(*d*) There are generally four sets of shadow scales[3] two on the periphery of

---

[1] In the Herāt (figure 12) and the Shāh Hussain instruments the main divisions are not equal but proportional to the sun's declination.

[2] Figure 37 shows how the curve (i) is constructed for latitude 27 N. For each pair of signs radii marking the angle of the meridian altitude of the sun are drawn, and the points of intersection of these radii with the corresponding arcs of the signs are joined. To make the graph perfectly correct intermediate points must, of course, also be fixed. Figure 11 shows the unequal hour curves. According to Delambre (*Astronomie du Moyen Age*, p. 243) this is first described in a small work by Sacrobosco (circa A.D. 1250). It occurs on many old astrolabes, but it gives only roughly approximate results. To construct these hour lines the arc of the quadrant is divided into six equal parts and a semi-circle erected on one of the bounding radii. This is the sixth hour line and the others are arcs of circles of which the centres are on the same radius, but at points equidistant from the centre of the quadrant and from the successive points of division on the arc of the quadrant.

[3] Figure 23 shows how these scales are constructed. There are generally two kinds of scales—one for a 7-unit gnomon and the other for a 12-unit gnomon. The semi-circle *a c e* is bisected at *c*, and *a c* and *c e* are bisected at *b* and *d*. The line *bd*, which is bisected at c^1, is the basis for both scales: the part bc^1 is divided into 12 equal divisions and c^1d into 7 equal divisions and these graduations are continued on their respective sides as far as is convenient. Lines joining the points of division with the centre cut the arcs and the small central rectangle and form on them the shadow scales. On most astrolabes the 12-unit scales are on *a b c* and *k g*, etc.: and the 7-unit scales on *c e* and *i l*. The Zarqālī astrolabe has, however, in place of the central rectangle, the smaller semi-circle, *fj* graduated. Variations occur, *e.g.*, the basis line (bc^1) may be placed vertically through *b* and then graduated.

the lower half of the disc and two in the central rectangle. In the Zarqālī instrument, however, the latter are replaced by a pair of central circular shadow scales (figure 19). Given the length of the shadow in terms of its gnomon—by the aid of the shadow scale and the alhidade the degree of altitude of the sun and the time can be found.

(*e*) In most of the instruments the inner semi-circles of the lower half of the disc give lists of the signs, manzils, terms, faces, etc. These tables are generally for astrological purposes and they are explained in some detail in a note on astrology appended to this volume (appendix B).

Some of the special tables are of great interest. The following are taken from A, B, and C (figures 7, 8 and 12).

(*f*) A table of rising signs is given in the central rectangle of 'Jaipur B.'

Table of times of rising of the signs for the latitudes of certain cities in India.

	39	38	37	36	35	34	33	32	31	30	29	28	27	26	25	24	23	22	21	20	LATITUDES.	
	H. M.	H. M.	H. M.	H. M.	H. M.	H. M.	H. M.	H. M.	H. M.	H. M.	H. M.	H. M.	H. M.	H. M.	H. M.	H. M.	H. M.	H. M.	H. M.	H. M.	SIGNS.	
t_1	1 14	1 15	1 16	1 18	1 19	1 20	1 21	1 22	1 24	1 25	1 26	1 27	1 28	1 29	1 30	1 31	1 32	1 33	1 34	1 35	Pisces	Aries.
t_2	1 28	1 29	1 31	1 32	1 33	1 34	1 35	1 36	1 37	1 38	1 38	1 39	1 40	1 41	1 42	1 43	1 44	1 44	1 45	1 46	Aquarius	Taurus.
t_3	1 56	1 57	1 57	1 57	1 57	1 58	1 58	1 58	1 59	2 0	2 0	2 0	2 1	2 1	2 2	2 2	2 2	2 3	2 3	2 3	Capricornus	Gemini.
t_4	2 22	2 21	2 21	2 20	2 20	2 20	2 19	2 19	2 18	2 18	2 17	2 17	2 17	2 16	2 16	2 16	2 15	2 15	2 15	2 14	Sagittarius	Cancer.
t_5	2 31	2 30	2 29	2 28	2 27	2 26	2 25	2 24	2 22	2 22	2 21	2 20	2 19	2 18	2 17	2 17	2 16	2 16	2 14	2 14	Scorpio	Leo.
t_6	2 30	2 28	2 27	2 26	2 2	2 23	2 22	2 21	2 20	2 19	2 18	2 16	2 15	2 14	2 13	2 12	2 11	2 10	2 10	2 9	Libra .	Virgo.

This table and the rule on which it is based played a very important part in mediæval astronomy and astrology.[1]

(*g*) **Trigons.**—On the right of the lower half of 'Jaipur B' (figure 8) and in the centre of C (figure 12) is a table showing the regents of the trigons, etc.

Nature of the Trigons and their Regents or Lords.

Nature of the Triplicities.	Fiery.			Earthy.			Airy.			Watery.		
Day Lords .	Sun .	Jupiter .	Saturn .	Venus .	Moon .	Mars .	Saturn .	Mercury	Jupiter .	Venus .	Mars .	Moon.
Triplicities .	Aries .	Leo .	Sagittarius	Taurus .	Virgo .	Capricornus.	Gemini .	Libra .	Aquarius	Cancer .	Scorpio .	Pisces.
Night Lords .	Jupiter .	Sun .	Saturn	Moon .	Venus .	Mars .	Mercury	Saturn .	Jupiter .	Mars .	Venus .	Moon.

The triplicities or trigons are groups of three signs each of which is situated 120 degrees from the other two. It will be noticed that Saturn, Mars, Jupiter and the Moon occur in their respective triplicites both as day and night regents and they are sometimes therefore termed 'common regents' (See Appendix B).

[1] See Book ii of the *Almagest*; al-Battānī *Opus Astronomicum* 2nd *Part*, p. 65f; the *Sūrya Siddhānta* ii, 60f. etc.; also the note on astrology (Appendix B). From the formula $\sin \alpha_1 = \tan \phi . \tan \delta$, where ϕ is the latitude and δ the declination the so-called *ascensional differences* are calculated; then from $\sin b_1 = \frac{\sin 30^\circ . \cos \omega}{\cos \delta_1}$, $\sin (b_1 + b_2) = \frac{\sin 60^\circ . \cos \omega}{\cos \delta_2}$, $\sin (b_1 + b_2 + b_3) = 1$ the values of b are calculated; and finally $t_1 = b_1 - a_1$, $t_2 = b_2 - a_2 + a_1$, $t_3 = b_3 - a_3 + a_2$, $t_4 = b_4 - a_4 + a_3 = b_3 + a_3 - a_2$ and so on since $b_4 = b_3$, $b_5 = b_2$, $b_6 = b_1$, and $a_4 = a_2$, $a_5 = a_1$.

(*h*) **Climates.**—The table of climates occurs only on Jaipur A (*figure 7*) towards the bottom of the lower half of the disc.

Table of climates.

CLIMATES.	FIRST.	SECOND.	THIRD.	FOURTH.	FIFTH.	SIXTH.	SEVENTH.
Beginnings (*b*) and middles (*m*)	b m	b m	b m	b m	b m	b m	b m
Latitudes . .	12^{43} 16^{44}	20^{21} 24^{10}	27^{28} 30^{00}	33^{43} 36^{20}	39^{1} 41^{21}	43^{30} 45^{28}	47^{30} 48^{59}
Hours . .	12^{45} 13^{0}	13^{15} 13^{30}	13^{45} 14^{0}	14^{15} 14^{30}	14^{45} 15^{0}	15^{15} 15^{30}	15^{45} 16^{0}

This topic of 'Climates' recalls a most interesting chapter in the history of civilization. It exercised the attention of such astronomers and astrologers as Eudoxus, Eratosthenes, Hipparchus, Manilius, Ptolemy, Dorothea of Sidon, etc., etc. The subject presented difficulties. The number of climates assumed varied, but generally a chorographic system, which applied the seven planets to the seven zones or climates, prevailed. Also, according to Paul of Alexandria, "each sign corresponds to a climate or parallel, and by virtue of its Trigon to each quarter." For the mathematicians, the problem was to find a progression corresponding to ascensional differences. They took the length of the day as the measure[1], and progressed from one climate to another by half hour steps. (See appendix C.)

(*i*) **The year.**—The rectangular table to the left in Jaipur A (*figure 7*) shows multiples of the differences between the approximately correct length of the tropical year and 365 days, thus:

67^{58}	340^{25}	252^{52}	165^{19}	77^{45}	350^{12}	262^{39}	175^{6}	87^{33}
9	8	7	6	5	4	3	2	1
90	80	70	60	50	40	30	20	10
319^{39}	164^{8}	$0\ 8^{37}$	213^{6}	57^{35}	262^{4}	106^{33}	11^{2}	155^{31}

The table gives $n(87^\circ\ 33'\ 6'')-a.360^\circ$ where n ranges from 1 to 9 and from 10 to 90 and a is a whole number.[2] Now $87^\circ\ 33'\ 6''$, expressed in time, is 5 hours 50 minutes 12·4 seconds, and the length of the tropical year was supposed to be 365 days 5 hours 50 minutes 12·4 seconds.[3]

---

[1] The measure of the longest day is $\frac{180^\circ + 2h}{15}$ where $\sin h = \tan \phi \cdot \tan \omega$.

[2] For example $106^\circ\ 33' = 30\ (87^\circ\ 33'\ 6'')-a.360 = 2626^\circ\ 33' - a.360^\circ = 7.360^\circ + 106^\circ\ 33' - a.360^\circ$ and $a=7$.

[3] A British Museum astrolabe dated A. H. 1070 (=A.D. 1659-60), by Mahammed Mokim of Lahore, gives the same table.

Al-Battānī gives 86° 36′ and Nallino gives the following note:" Ḥabash in suo astronomiæ libro narrat partem excedentem revolutionis (*fazl al-daur*), scilicet quantitatem (gradibus expressam) qua Solis totus ambitus 365 dies excedit, inventam esse ab astronomis khalifae al-Ma'mūn:—

in specula al-Shammāsiyyah. . . . . 86° 43ᴵ 39ᴵᴵ 36ᴵᴵᴵ 47ᴵⱽ
Damasci in urbe . . . . 86° 41ᴵ 25ᴵᴵ 14ᴵᴵᴵ 9ᴵⱽ
a Yaḥy'a ibn Abī Manṣūr . . . . 86° 35ᴵ 13ᴵᴵ 46ᴵᴵᴵ

In tabulis suis Ḥabash primam quantitatem recepit, rotunde scribens 86° 43ᴵ 39ᴵᴵ 37ᴵᴵᴵ, vel, gradibus in tempus conversis, 5ʰ 46ᵐ 54ˢ 38ᵗ etc." *Opus Astronomicum* I. pp. 42 & 211.

The present length of the tropical year expressed in mean time is 365 days 5 hours 48 minutes 45·5 seconds or 365·2422 days nearly.

Fig. 17. TABLET FOR LATITUDE 32.

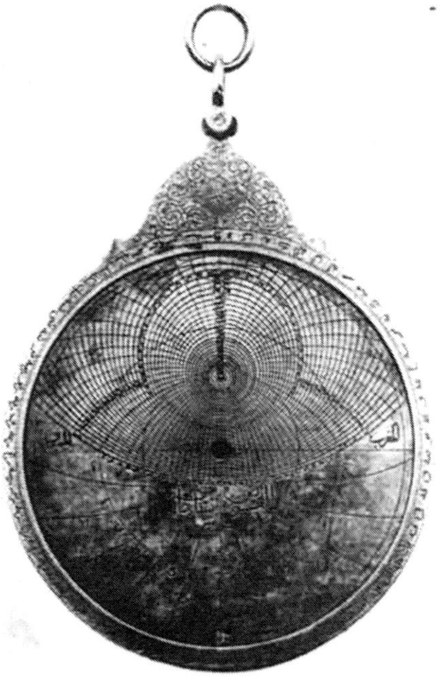

Fig. 18. TABLET FOR LATITUDE 28.

Fig. 19. ZARQALI ASTROLABE (OBVERSE).

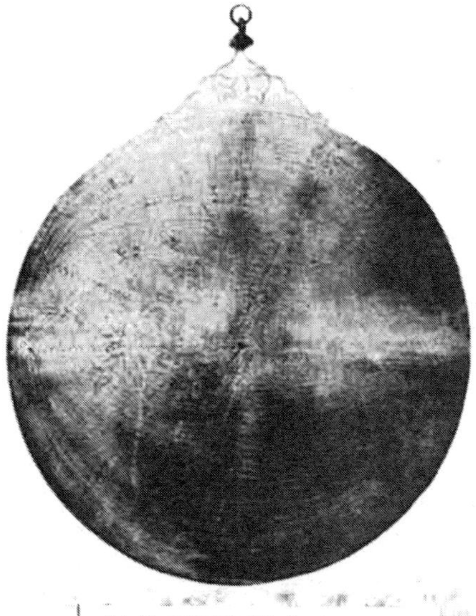

Fig. 20. ZARQALI ASTROLABE (REVERSE).

Photogravure. Survey of India Offices, Calcutta, 1917.

14. **The face.**—(*Wajh*) *or* Venter.—The inside surface (*venter*) of most of the astrolabes is engraved with a list of cities and their latitudes and longitudes and sometimes their position relative to Mecca.[1] The last is indicated by the *inhirāf* or 'inclination' and the *masāfat* or 'distance' and *jihat* (side or point of the compass). The inhirāf is the arc of the horizon intercepted between the meridian of the place and the vertical circle passing through the zenith of Mecca.[2] The Masāfat is the distance of Mecca measured along a great circle and the jihat is the quarter of the horizon in which Mecca lies.

Jaipur B (figure 13) gives the longitude and latitude of 210 places and also the districts in which they are situated but does not give the inhirāf etc.; Jaipur D gives 36 towns; the Herāt astrolabe (C) gives 44 places with the inhirāf and jihat; the Shāh Hussain instrument gives 103 places with latitude, longitude, inhirāf, masāfat and jihat; the Lahore astrolabe gives also 36 places with latitudes and longitudes only. A selection from these astrolabe gazetteers is given in an appendix (pp. 127-8).

The longitude is in all cases reckoned from the 'Fortunate Islands.' Compared with a modern atlas the differences for a few selected[3] places are—

	Longitude difference.	Latitude difference.
Marāghah	35° 43′	0° 1′
Baghdād	35° 22′	0° 4′
Shirāz	35° 20′	0° 6′
Nisābūr	33° 50′	0° 13′
Yezd	34° 30′	0° 22′
Iṣfahān	34° 56′	0° 14′

The difference in latitude may be taken as some criterion of the accuracy of the determinations, but it must be remembered that the precise localities of the observations (old and new) are not known.[4]

The longitude differences point to some place about 35 degrees west of Greenwich as the point of origin. The zero meridian therefore passed through the Azores. In this matter the Muhammadans copied the Greeks who fixed upon the 'Fortunate Isles,' possibly, as the western end of the world. These 'Fortunate Isles' were originally imaginary islands where the souls of the good were made happy, but later the name became attached to the Canary Islands. In the *Ain-i-Akbari*[5] we read: "The Greeks commence their reckoning from Khalidat, which are six islands in the western ocean, which in ancient times were inhabited, but now are inundated, etc."

---

[1] See the excellent little book *Paraphrase de l'astrolabe* written in A.D. 1555 by Jaques Focard de Montpeltier, who terms the venter 'Miroer du Monde' and gives on it an actual map of the world (i. 136-7).

[2] The *angulus positionis* of the old geographers. See al-Battānī, i. 136-7; L. A. Sédillot's *Mémoire*, 97f.; the *Nuzat al-Qutūb* (Ed. G. le Strange) p. 26; &c. &c.

[3] The values given on all the instruments examined for these six places are the same: at Marāghah, Baghdād and Nisābūr were important observatories.

[4] One second of longitude at Delhi is roughly equivalent to about 30 yards; very roughly a mile to a minute of arc. (At latitude 30° N. one degree of longitude = 96489 metres = 59·97 miles.)

[5] Ed. Gladwin, ii 351.

The longitude values are somewhat irregular and illustrate the inherent difficulty in the determination. The errors for Jerusalem and Cairo and Mecca seem to be traditional. The list of names on 'Jaipur B' (figure 13) is interesting as being copied from Ulugh Beg's table.[1] The number of towns is the same, but the astrolabe designer left out some of the western places [*e.g.*, those in the country of Rūm and some of those of Shām (Syria)], and added some extra towns for India. Otherwise the names, the order and the latitudes and longitudes are the same.

15. **The Alhidade** (*'iḍādah*) or Sighter.—In all old oriental astrolabes the alhidade is of the type attached to Jaipur A and shown in figure 7. It is fixed on the centre pin so that its graduated edge lies on a diameter of the circle. Half of the bevelled edge (the right upper edge in figure 7) is divided into 60 equal divisions, every third division being numbered. The left upper edge is divided into six equal divisions, corresponding to the divisions for the signs of the zodiac in the south-west quadrant, and each division is marked with its two proper signs and divided into 15 parts. The right lower edge is divided into six divisions numbered 1 and 12, 2 and 11, etc. Near to each end is fixed a sighting tablet, each having two holes, the upper pair generally being the larger. On modern Hindu instruments is another type of revolving index. It has no sighters and is the length of the radius only. In mediaeval European instruments we often find such a marker or 'label,' as it was called,[2] used on the front of the astrolabe. (Figure 24.)

The drawings A, B, C (fig. 27) are taken from Morley's work.[3] The last C, is from Focard [4] and the others, A and B, Morley copied from Ritter.[5]

[1] See L. P. E. A. Sédillot, *Prolégomènes des tables astronomiques d' Oloug-Beg.* 1853, p. 257 *f.* Compare also al-Battānī's list (*Opus Astronomicum* ii. 33—54).

[2] *Ostensor, Ponella.*

[3] A is fig. 34, B fig. 36 and C fig. 32 of Plate XXXI of Morley.

[4] *Paraphrase de l'astrolabe*, Lyon, 1555.

[5] *Astrolabium*, Franc. Ritteri, Nurnberg. 1613.

Chapter IV.—THE ZARQĀLĪ ASTROLABE.

16. The instrument shown in figures 19 and 20 is dated the 23rd year of the reign of Aurangzeb and A. H. 1091 (A.D. 1680), and was made at Delhi for Nawab Iftikhār Khān of Jaunpur[1] by a certain Ziā al-Dīn b. Mullā Qāsim Muḥammad b. Ḥāfiẓ Īsā b. Allah Dād, Humāyūnī, asṭūrlāb maker of Lahore.[2] The instrument is now at Jaipur. It is labelled '*Yantra Jara Kālī sarva deśī.*' It is two feet in diameter and is made of brass and consists of one disc only engraved on both sides and is described in the inscription as a '*Zarqālī astrolabe*'[3] and as a 'single leaf' instrument.

The invention of this instrument is usually attributed to Ibrāhīm b. Yaḥyā al-Naqqāsh (the engraver), born at Cordova, and who lived from about A.D. 1029 to 1087.[4] He was known as al-Zarqālī (Arzachel) and his instrument was called *al-ṣafīḥat al-zarqāliya* 'the tablet of al-Zarqālī'[4] and was famous in mediaeval Europe under the name *Saphaea Arzachelis.*

17. The characteristic part of the instrument is engraved on the reverse. The ordinary astrolabe was considered inconvenient, inasmuch as for every new latitude an additional tablet was required. Al-Zarqālī tried to remove this difficulty by substituting a horizontal projection for the usual polar projection. He took as his centre of projection one of the equinoctial points, and made the solstitial colure (*i.e.*, the great circle passing through the solstitial points and the poles of the equator) the plane of projection.[5]

The projections of the two celestial hemispheres exactly coincide, and the scheme can be used for any geographical latitude. By the aid of the index and sighter most of the results given by the ordinary astrolabe can be obtained. The index of the Jaipur instrument consists of cross bars (four arms at right angles).

---

[1] The reading of this word is uncertain, but Iftikhār Khān was possibly Sulṭān Husain, who in the first year of 'Alamgir was given the title 'Iftikhār Khān' and who was Faujdār of Jaunpur where he died in A. H. 1092 (= A.D. 1681-2).

[2] This appears to be the maker also of the very fine 'complete' astrolabe (Jaipur B, figures 6 and 8) made in A.D. 1657).

[3] Al-Zarqālī called it *al-'Abbādiya* in honour of al-Mu'tamid b. 'Abbād, king of Seville (A.D. 1068-1091). See C. Nallino. *Encyclopaedia of Islām* p. 602.

[4] Suter (p. 109), who also records (p. 163) that one Aḥmad b. M. b. 'Othmān al-Azdī, who died at Morocco about A.D. 1340, also wrote on the use of the *Sabāri* and *Zarqālī Ṣafīḥa*. For a brief description of the instrument see also L. A. Sédillot's *Mémoire*, p. 184 and fig. 95.

[5] In figure 25, PP^1, are the poles of the equator, K^1 is a pole of the ecliptic, and Pr. P' ♎ is the equinoctial colure. The centre of projection is ♎ and $p\ c\ q\ p^1$ is a half of the projected solstitial colure. In the figure circles of declination, etc., are drawn for angles of 30°, 45°, and 60°. D_1D_1, D_2D_2 and D_3D_3 are circles of declination, pA_1p^1, pA_2p^1, pA_3p^1 are circles marking right ascension; similarly L_1L_1, L_2L_2, L_3L_3 are circles of latitude, and l_1K_1, l_2k_1, l_3k_1 circles of longitude. On the instrument itself circles of declination and latitude are drawn for every degree, and those for right ascension and longitude for every three degrees. The latitudes and declination circles are numbered from the centre to the poles; the ecliptic is marked with the signs of the zodiac from O to F (figure 21), and back again from F to O, and similarly along the other half of the ecliptic line, and each sign is graduated for every three degrees; the graduations on the equator commence at D and proceed on the north side of the line to B which is numbered 180, and then the graduations are continued on the south side of the line ending up at D, which is this time numbered 360.

The projection was used for mapping out both the celestial hemispheres. For example we have (fig. 21) the stars *Simāk-Rāmiḥ* (Arcturus) and *Fam al-Faras* (Enif) placed near each other something like this

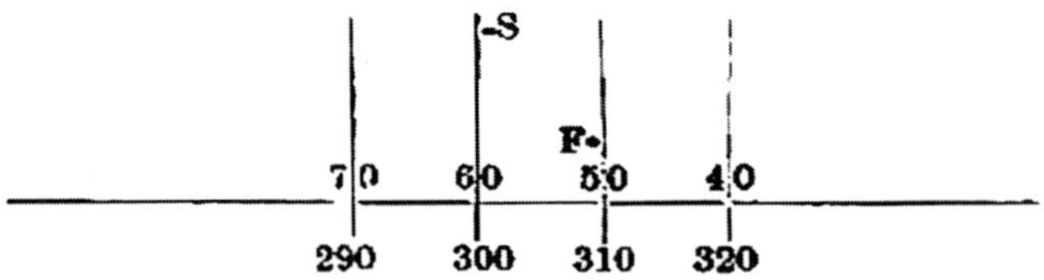

Their right ascensions may therefore be read

F	52	or	308
S	59	or	301

But, as the scale on the instrument commences at the winter solstice,[1] and the first point of Aries is reckoned as 90 instead of 0 we must deduct 90 we then have

F	—38	or	218
S	—31	or	211

According to Flamsteed, F was approximately 322°, which is the same as —38° (360°—322=38) and S was approximately 210½°.

In Appendix A4 the names and approximate positions of certain stars on the instrument are compared with the positions as given by Flamsteed.

18. There are also the names of several stars written in the Devanāgarī character, *e.g.*, *Marīchi* with R. A. 204° and declination 51°N., and *Pulaha* with 19½° or 160½° and 63 N. The former is usually identified with η *Ursæ Majoris*, for which Flamsteed gives R. A. 203° 50′ and declination 50°53′; and *Pulaha* may be α *Ursæ Majoris*, for which Flamsteed gives 161° 2½′ and 63° 25′.

Also the names of a number of towns are given, *i.e.*, the celestial map is also used for geographical purposes. The axis CA appears to have been taken as zero longitude, and Baghdād, or some place with nearly the same longitude, appears to have been considered as situated on this axis.

[1] al-Hasan b. 'Alī b. 'Omar al-Marrākoshī, Abū 'Alī (fl. A.D. 1262) appears to have been the first to employ Right ascensions, which he reckoned from the first point of Capricornus. See E. B. Knobel, *The Chronology of Star Catalogues*, p. 14.

Some of the towns and their positions are given below :—

Name of Town.	Longitude with reference to C A in figure 21.	Latitude with reference to D B in figure 21.	Longitude with reference to Baghdād as given on Jaipur B.	Latitude as given on Jaipur B.
	°	°	° ′	° ′
Halb (Aleppo)	—8	+35¾	—7 50	+35 30
Tus	+13	+37	+12 30	+37 0
Kābul	+24½	+34	+24 40	+34 7
Jahānābād (Delhi)	+34	+28½	+33 35	+28 30
Lahore	+29½	+31½	+29 20	+31 50

19. The obverse of the disc (figure 19) contains ~~the following~~ scales and tables[1], which are, more or less, common to all astrolabes. Reading from the circumference towards the centre ~~they are :—~~

(*a*) The two upper quadrants ~~are~~ graduated for every three degrees (numbered in the abjad notation) also in degrees numbered in Arabic numerals from 1 to 90, and in one-sixths of degrees, or every twelve minutes.

(*b*) The periphery of the lower quadrants is graduated by shadow scales—on the left a 'twelve scale', and on the right a 'seven scale' (See p. 22).

(*c*) The next complete annulus contains the signs of the zodiac, which are accompanied by graduations down to intervals of twelve minutes.

(*d*) Next are the *manzils* or 'mansions of the moon'.

(*e*) The planets—twelve to each sign—with graduations for every 2½ degrees.

(*f*) The planets—nine to each sign—with graduation, for every 3° 20′.

(*g*) The planets—five to each sign—with their limits or terms indicated.

(*h*) The planets—seven to each sign—at intervals of 4⅔ degrees.

(*i*) The planets—three to each sign. These are the 'faces' of the particular sign.

(*j*) Again three planets to each sign.

(*k*) Another pair of shadow scales.

(*l*) Separated from the others by the smaller shadow scales (*k*) are the names of the European months, with a scale showing the days of each month, etc. The instrument was made in A. D. 1680 and correctly indicates that spring commenced on March 10th.

20. The central part of the disc consists of a projection of a sphere and a table of sines.[2] These are illustrated in figure 22, where the quadrant OAB forms the table of sines. The arc AB is divided into degrees, and, from every point of division lines

[1] Of these, c to j are shown in the appendix on astrology (p. 124).

[2] On an astrolabe made at Seville in A.H. 609 (A.D. 1211-12) similar constructions are found. See the articles by MM. Sauvaire and Pailhade, *Journal asiatique*, 1893, 9 série, i, pp. 6f. and 185f.

are drawn perpendicular to OA. The radius OB is divided into 60 equal parts, and lines are drawn parallel to OA. In the diagram *sin* 40° reads 39°, *i.e.*, $\frac{39}{60}$ or ·65, but the instrument itself is more accurate than this. The radius OD is divided into sixty equal parts, and, through each point of division, circles, also passing through the points A and C, are drawn. These arcs are orthogonal projections of great circles inclined to the meridian C'A' of the sphere ABCD. For example, the arc passing through the division numbered 50 represents a circle on the sphere inclined to the meridian at an angle ϕ, such that sin ϕ $=\frac{50}{60}$ (=·835). Now the arc 50 in the quadrant CB touches the horizontal line 50 which cuts the arc BA at 56½, therefore ϕ =56½ [actually sin 56½=·8339].

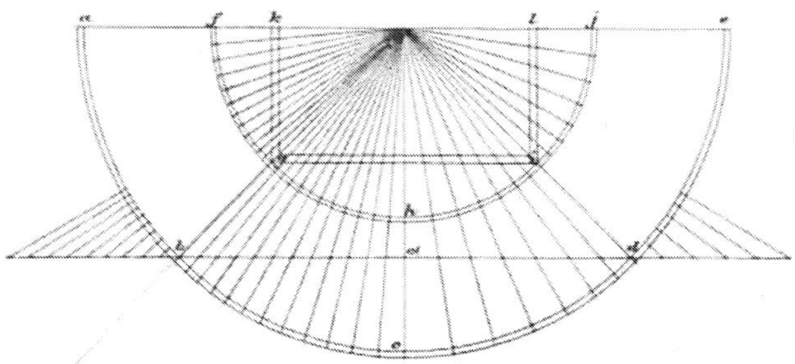

FIG. 23. SHADOW SCALES.

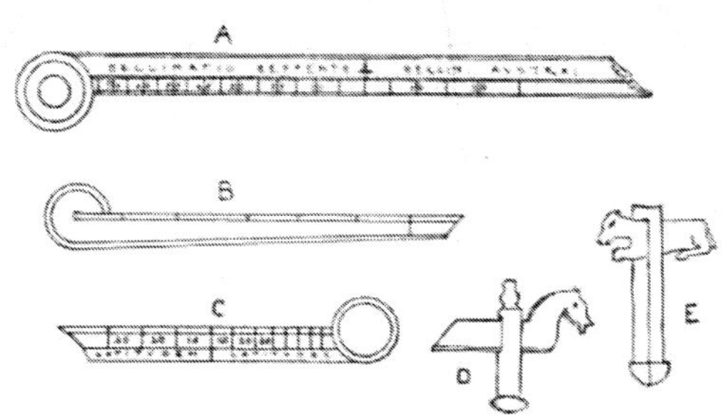

FIG. 24. RULERS AND WEDGES AFTER MORLEY.

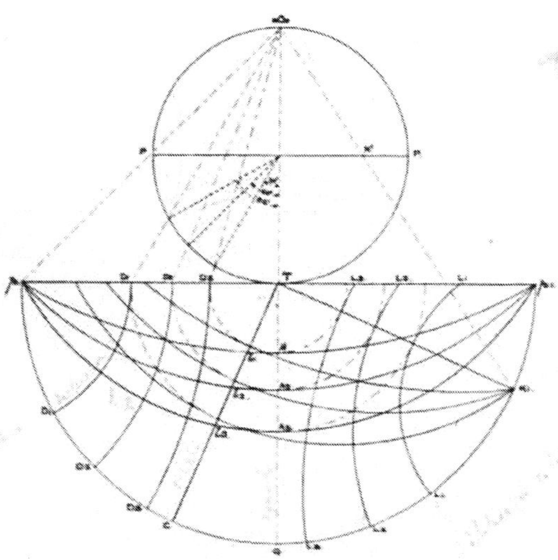

FIG. 25. PROJECTION FOR FIGURE 19.

CHAPTER V.—HINDU METAL INSTRUMENTS.

21. Hindu astrolabes are mostly of modern workmanship and of modern pattern; but at Jaipur there is a Hindu copy of a Persian astrolabe that is of interest. It is of inferior workmanship and was, apparently, never properly completed. It is a *thulthi* or tripartite instrument and has two tablets, for latitudes 27° and 72° N., and a tablet of celestial latitudes and longitudes. On the 'ankabūt the ecliptic is graduated at intervals of 6 degrees (figure 26); and on most of the points no names are engraved. On the back of the instrument (figure 27) is the usual table of sines and declination graphs for 27° and 28° 39′. On the lower half are the usual shadow scales but nothing else.

Morley describes two other Hindu astrolabes one belonging to the Royal Asiatic Society and the other to the India Office.[1] (1) That belonging to the Royal Asiatic Society is a bipartite instrument and appears to contain one disc only for latitude 24°N. The 'ankabūt has 23 points with the names of stars engraved thereon. The back has the table of sines and the shadow scales. (2) The India Office instrument is said to be of poor workmanship. It is a sexpartite instrument, 3 inches in diameter. Within the *umm* is a table of 16 Indian cities, with latitudes and longitudes, the latter reckoned from the 'Fortunate Isles,' *e.g.*—

	Latitude.	Longtitude.
Jayanpur	26° 36′	109° 6′
Ujjeyani	23° 30′	110° 50′
Delhi	29° 0′	113° 0′
Benares	26° 15′	117° 20′

There are seven tablets—six for latitudes 0,° 17°, 18°, 20°, 21°, 23°, 24°, 26°, 27°, 29°, 32′ and 72° and one with usual horizons on one side and the 'ankabūt co-ordinates on the other. On the back is a set of tables termed *paramakrānti.*

In 1790 R. Burrow related that he "Compared an Astrolabe in the Nagry character (brought by Dr. Mackinon from Jaynagur) with Chaucer's description, and found them to agree most minutely, "even the centre pin which Chaucer calls 'the horse' has a horse's head upon it in the instrument."[2]

The only other ordinary astrolabe of Hindu make and of any age known to me is R in the list on p. 17. It is engraved in Devanāgarī character and is of very crude workmanship as compared with A, B, C, D and E. It is of copper. of 7 inches diameter, and contains two tablets, besides the ecliptic tablet ('ankabūt). The venter is plain, while the back has only the central rectangular shadow scales, the *sinus quadrantus* ruled into 30 equal divisions, and the declination quadrant divided into even spaces by 15 arcs. The two tablets are tripartite and besides the almucantarats and azimuth lines have the equal and temporal hour lines. They are inscribed thus:—

(a_1) Longest day 33—30
AVANTI.

(a_2) Longest day 36—24
Latitude 37.

[1] The India Office instruments are now in the Indian section at the South Kensington Museum.

[2] *Asiatic Researches,* 1790, Vol. ii, p. 483.

Latitude 22, Shadow 5.
Hypotenuse 13.
(b_1) Longest day 33—50
AMADĀVĀD.
Latitude 23.
Shadow 5—6.
Hypotenuse 13—3.

Shadow 9.
Hypotenuse 15.
(b_2) Tablet of horizon without any numbers or inscriptions.

The latitudes are here given in four different ways (*i*) in degrees, (*ii*) in length of longest day (in ghaṭis and palas), (*iii*) in length of the equinoctial shadow[1] or $\tan\phi$, (*iv*) by the hypotenuse of the equinoctial shadow or $\sin\phi$.

These may be expressed thus:—

Values on the Instrument.									Calculated values.				
Lat.	Longest day.						tan φ	Sin φ	Longest day.			tan φ	sin φ
°	g.	p.		h.	m.	s.			h.	m.	s.		
22	33	30	=	13	24	0	$\frac{5}{12}$ = ·417	$\frac{5}{13}$ = ·385	13	22	52	·404	·375
23	33	50	=	13	32	0	$\frac{5°6'}{12}$ = ·439	$\frac{5°6'}{13°3'}$ = ·391	13	27	16	·424	·391
37	36	24	=	14	33	36	$\frac{9}{12}$ = ·750	$\frac{9}{15}$ = ·600	14	36	48	·754	·588

22. At Jaipur are two large single disc astrolabes 7 feet in diameter—one made of some sixty sheets of iron rivetted together (figure 28) and the other of brass patched up with lead (figure 29). From the iron instrument the graduations have disappeared. The brass instrument is *tamm* (complete) for latitude 27°N. It has an ecliptic circle and a tube sighter of modern workmanship. These two instruments may possibly be of the original metal instruments referred to by Jai Singh (See p. 12). If so they were probably brought from Delhi.

23. *The Unnatāṁsa Yantra* is possibly another of Jai Singh's original instruments. It is a graduated brass circle 17½ feet in diameter suspended so as to revolve around a vertical axis. Jai Singh speaks of an instrument "*Zat-al-Halqa* (consisting of a ring) of brass, in diameter three *gaz* of the measure now in use," (See page 12) but the *Zat al-ḥalqa* is ordinarily an armillary sphere.

The Chakra Yantra (circle instrument) is an equatorial. There are two at Jaipur each 6 feet in diameter (figure 57) and one at Benares, 3 feet 7 inches in diameter. The Chakra Yantra is fixed so as to revolve about an axis parallel to the earth's axis. At the southern end of the axis of the instrument is a separate graduated circle fixed on the supporting pillar. The axis carries a pointer which indicates the hour angle on the fixed circle; and the main movable circle carries an index and sighter (figure 68).

The Krānti Vṛitti Yantra ('Ecliptic instrument') found at Jaipur only is quite a modern instrument but is said to have been made according to

[1] The gnomon is supposed to be twelve units or 720 minutes in length.

PLAN

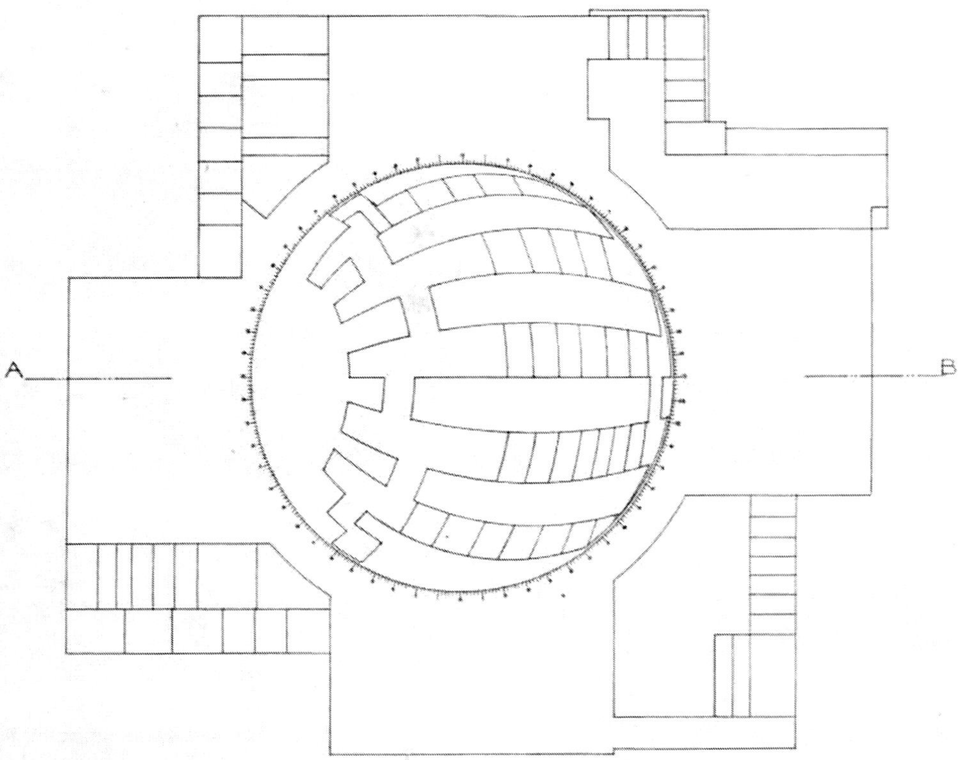

SECTION ON A.B.

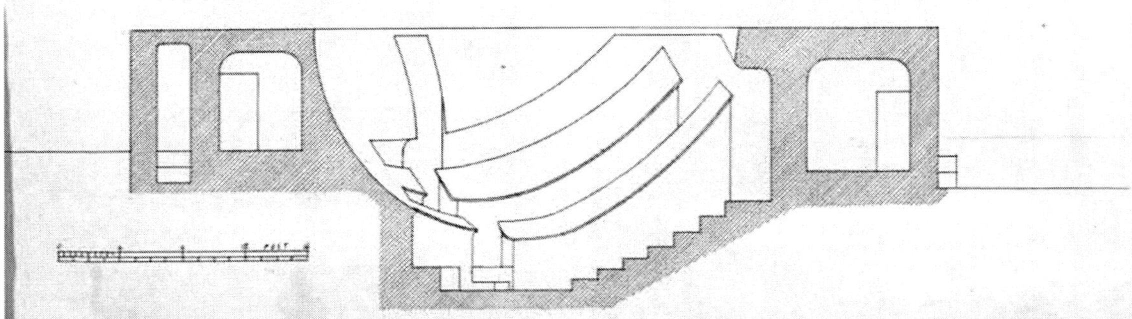

THE JAI PRAKAS, DELHI.

Fig. 26. OBVERSE OF HINDU ASTROLABE (JAIPUR III).

Fig. 27. REVERSE OF HINDU ASTROLABE (JAIPUR III).

Fig. 28. IRON ASTROLABE (JAIPUR II).

Fig. 29. BRASS ASTROLABE (JAIPUR I).

Photo-etching. Survey of India Offices, Calcutta, 1917.

Jagganāth's instructions; and there still is at Jaipur the old masonry work for a much larger instrument of the same type. The Krānti Vṛitti Yantra is used for direct measurements of celestial latitude and longitude. It consists of two brass circles (figure 58), pivoted so that one always moves in the plane of the equator and the other in the plane of the ecliptic. It is more suitable for demonstration purposes than for actual observation. (See page 51.) This is the *Torquetum* of Regiomontanus (1434-1476), which was rejected by Tycho Brahe as a clumsy instrument.[1]

24. At Jaipur I was shown a modern Hindu astrolabe or *Yantra Rāj*, dated Samvat 1799 (=A.D. 1877). It is a single disc *thulthi* or tripartite instrument or latitude 27° N, with an ecliptic circle and ruler. On the obverse of the disc are engraved azimuth lines for 30°, 60° and 90°, and also the temporal hour lines and also the names of the following stars:—

Name on instrument.	Modern Name.	Name on instrument.	Modern Name.
1 Samudra pakśi . .	? *Ceti.*	9 Chitra	α *Virginis.*
2 Rohiṇī	α *Tauri.*	10 Svātī . . .	α *Boötis.*
3 Ārdra	α *Orionis.*	11 Anurādhā . . .	α *Scorpii.*
4 Lubdhaka . . .	α *Canis Majoris.*	12 Abhijit . . .	α *Lyræ.*
5 Pushya	δ *Cancri.*	13 Śravaṇa . . .	α *Aquilæ.*
6 Maghā	α *Leonis.*	14 Satajiva . . .	λ *Aquilæ.*
7 ?	..	15 Pūrvabhādrapadā .	β *Pegasi.*
8 Hastā	δ Corvi.	..	..

The ruler is of the same type as those employed on the face of certain mediæval European instruments (see p. 26).

25. The Dhruva Bhrama Yantra or 'Circumpolar instrument' is another modern Hindu instrument of rather crude workmanship. It consists of a square plate, with a slit near to and parallel to one edge, and a freely revolving weighted index with four pointers. If the plate is held vertically in such a position that the Pole star and the star *Markati* (Kochab or *β Ursæ Minoris*) are in line with the slit, then the pointer marked *Ghaṭi* will indicate sideral time in *ghaṭis*.[2] The other pointers indicate the rising sign, the sign on the meridian, and the rising, meridian and setting nakshatras.[3] The back of the instrument is marked *Turīya Yantra* (quadrant instrument), and consists of a hinged rod, two sighting rings on the edge parallel to the slit and a graduated quadrant consisting of eleven scales. When the sun shines through the sighting ring the index shows the altitude and the time. Also a list of the 28 *nakshatras* (initials only), starting with Aśvini and proceeding in the usual order[3], is given, and to each asterism is attached a number varying from 12¾ to 18¼.

[1] See R. Wolf, *Geschichte der Astronomie*, p. 161 and J. L. E. Dreyer, *Tycho Brahe* p. 317.

[2] The Hindus reckon their sideral time from the rising of the vernal equinox and hence it differs from European time by 6 hours.

[3] See *Garrett*, p. 62 and plate X.

6. **Other instruments.**—Some time before 1839, Rāja Rām Singh of Kotah presented to the Government of India an instrument similar to the *Dhruva Bhrama Yantra* described above. The Raja's instrument was of massive silver[1] and was made in A.D. 1834. On the reverse is the 'sine quadrant' usually found on astrolabes (see page 22).

The armillary sphere referred to by Tod (see p. 16) is still in existence, and is a very elaborate affair, although not of much practical use.[2]

At the Lahore exihibition of 1864 were several astronomical instruments of interest—particularly some astrolabes from Kapūrthala. The list of instruments—drawn up by a Hindu astrologer—is curious and valuable.[3] One of the entries is—" Yanti Rāj—the usturlāb of the Yunānī."

[1] *Journal of the Asiatic Society of Bengal*, 1839, p. 831f.

[2] I am indebted to Mrs. Borough Copley of Kotah for this confirmation of Tod's statement and for a photograph of the armillary sphere.

[3] See B. H. Baden Powell's *Hand Book of the Manufactures and Arts of the Punjab*, 1872, p. 259f.

CHAPTER VI.—MASONRY INSTRUMENTS.

27. The masonry instruments, which vary in size from a few feet to 90 feet in height, are Jai Singh's chief work. It has already been related how Jai Singh discarded brass instruments and built massive masonry ones in their place. His reasons appeared to be, but were not altogether, sound. The brass instruments were, he said, faulty because of their mobility and size.[1] The axes became worn and the instruments untrue; the graduations were too small for fine measurements, etc. His remedy was to make large immovable instruments: but he thus stereotyped his designs and hindered futher improvements. The larger and more immobile an instrument is, the greater is the difficulty in making alterations and improvements. Jai Singh sacrificed facility for supposed accuracy.

Hunter states that Jai Singh himself devised the Samrāṭ Yantra, the Jai Prakāś and the Rām Yantra. These three instruments are indeed peculiar to Jai Singh's observatories and must be to some extent attributed to Jai Singh's personal ingenuity.[2] Jai Singh used other stone instruments, such as the mural quadrant and cylindrical dial but these were not mentioned specially in the preface, because they were common to many observatories. They are however mentioned in Jagannāth's introduction to the *Samrāṭ Siddhānta* (see page 3).

The masonry instruments are :—

(*a*) *Samrāṭ Yantra* at Delhi, Jaipur (2), Ujjain and Benares (2). Figures 34, 35, 43-46, 66 and Plate XV.

(*b*) *Jai Prakās* at Delhi and Jaipur. Figures 30, 32, 33 and Plate XVIII.

(*c*) *Rām Yantra* at Delhi and Jaipur. Figures 47, 48, 49, 59 and Plate XVII.

(*d*) *Digaṁśa Yantra* at Jaipur, Ujjain and Benares. Figures 63, 65 and Plates XXIV and XXVI.

(*e*) *Dakshiṇovṛitti Yantra* at Jaipur, Ujjain and Benares. Figures 56 and 62.

(*f*) *Nari-valaya Yantra* at Jaipur. Ujjain and Benares. Figures 53 and 65.

(*g*) *Vṛitti Shasṭāṁśaka* at Delhi and Jaipur.

(*h*) *Miśra Yantra* at Delhi. Figures 50, 51 and Plate XIX.

(*i*) *Rāśi Valaya* at Jaipur. Figures 54 and 55.

(*j*) *Kapāla* at Jaipur. Figure 31.

The last three of these instruments are possibly of later date than Jai Singh. They are mentioned in neither of the contemporary lists.

---

[1] The contrast with the procedure in Europe is interesting. The European scientist recognised the inevitability of error and took measures to counteract it (*e.g.*, with the micrometer, vernier, telescopic sights, etc., etc.) Even a modern theodolite as a useful astronomical instrument is worth more than all Jai Singh's large buildings. Possibly Jai Singh's power and wealth inclined him to move in a direction that could not lead to the desired end. See page 90.

[2] See page 86.

28. The **Samrāṭ Yantra** or 'Supreme instrument'[1] is, as its name implies, the most important. It is an equinoctial dial, consisting of a triangular gnomon with the hypotenuse parallel to the earth's axes, and on either side of the gnomon is a quadrant of a circle parallel to the plane of the equator. It is in principle one of the simplest 'equal hour' sundials.

In figures 34 and 35 AB is one edge of the gnomon, the angle ABC is equal to the latitude of the place, EF and GH are at right angles to AB, as also are DF, MH. If KL is the direction of the sun then the arc KG indicates he time before noon, and the angle HGL the declination or sun's angular distance from the equator. In the actual structure the considerable width of AA' and GE (each being over 9 feet at Jaipur) practically duplicates the instruments. Each edge of the quadrants is graduated in hours and minutes[2], as well as in degrees and each edge of the gnomon has two scales of tangents, one from H to B, and the other from F to A. In the figure $tan\ HGL=\frac{HL}{GH}$ and GH is the radius of the quadrant MKG.

The shape of the gnomon is generally a parallel trapezium as, in figure 34, ABB'C. In the same figure GE represents the position of the quadrants as they enter the gnomon. $HG=FE$ is the radius, and the lines radiating from E and G show the construction of the scales of tangents on the edge AB.

In the following list the examples of the Samrāṭ Yantra are enumerated and their dimensions (with reference to figure 34) are given.

SAMRĀṬ YANTRA.

	Height.		Base BC.	Hypotenuse AB	Radius GH = EF	Width of Quadrant GE	Angle ABC approximate.
	AC'	AC					
Delhi	68'	60' 4"	113' 6"	128' 6"	49' 6"	7' 7½"	28° 37'
Jaipur	89' 9"	75' 3"	146' 11"	174' 0"	49' 10"	9' 3¾"	26° 53'
		18' 6"	37' 0"	40' 8"	9' 1½"		
Ujjain	22' 0"	18' 6"	43' 6"	47' 6"	9' 1"		23° 10
Benares	22' 3½	16' 11½	35' 10"	39' 8½"	9' 1½"	5' 10"	25° 14'
	8' 3"	4' 9"	10' 0½"	11' 1½"	3' 2"	1' 9"	25° 19'

These dials give apparent solar time which varies from day to day, owing to (1) the eccentricity of the earth's orbit and its consequent more rapid angular motion in the winter (when it is nearer the sun) and its slower motion in summer; (2) the obliquity of the ecliptic.[3] Consequently a clock going regularly does not agree for long with solar time. In India there is another element of difference to consider due to the standard time being fixed for the

[1] Williams says the Arabic name was '*Kootoop bede* in Hindu *drooj*.'

[2] They were originally graduated in *ghaṭis* and *palas*.

[3] These are two principal causes but all other causes combined only alter the equation of time by a few seconds.

longitude of 82½° degree east of Greenwich or 5½ hours before Greenwich time. A table that will enable the observer to compare roughly the dial time with clock time is given in appendix C.

29. The **Jai Prakāś** is called by Jagannāth *sarva yantra śiromaṇi* 'the crest jewel of all instruments.' It is a hemisphere on the concave side of which are mapped out certain co-ordinates. Cross wires are stretched north to south and east to west and the shadow of the intersection of the wires falling on the surface of the hemisphere indicates the position of the sun in the heavens: other heavenly bodies can be observed direct by 'placing the eye' at the proper graduated point and observing the passage of the body across the point of intersection of the wires. For this purpose passages are cut into the hemisphere and the instrument is duplicated.

The construction of the instrument is seen in the plan and section shown in plate XVIII, and in figures 30, 32 and 33. Figure 38 shows a projection of a complete Jai Prakāś. The outer circle represents the horizon and is graduated in degrees. From the centre azimuth lines and altitude circles are drawn. (These are not all shown in the plate) The pole P is at a point on the meridian line, BD, at a distance from the point B equal to the latitude of the place (In the plate 28°37′ approximately). The equator, AEC, and tropics, *fff* (Capricorn), *ggg* (Cancer), and intermediate diurnal circles (not shown) are drawn. The equator cuts the meridian at a point, E, at a distance from the centre point equal to the latitude of the place (28°37′), and the other circles cut the meridian at distances 23½°, 20°12′ and 11°30′ on either side of the equator. Through the pole, hour circles Pa^1 Pa^2, Pa^3, etc., are drawn. The circles hh_1, ii_1, jj_1, kk are circles of the signs, and are such, that when the shadow falls on any one of them the corresponding sign is on the meridian. Two such circles cut each of the seven diurnal circles on the meridian and cut the neighbouring diurnal circles at the proper intervals. At Jaipur a similar instrument (figure 31) called *Kapāla* ('cup' or 'hemisphere') is so constructed as to show 'rising signs'. In this instrument the edge of the hemisphere corresponds not to the horizon but to the solstitial colure (*i.e.*, the circle passing through the poles and the solstial points), and thus is the *Jai Prakāś* turned through a right angle.[2]

The Jai Prakāś is found only at Delhi and Jaipur. The diameter of that at Delhi is 27 feet 5 inches and that at Jaipur 17 feet 10 inches. (See plates XVIII and XXI).

30. The **Rām Yantra** is the third of the stone instruments mentioned in the preface to the *Zīj Muhammad Shāhī* (page 13). The Paṇḍits say it was named after Rām Singh, a predecessor of Jai Singh's. According to Hunter the instrument was also known as *Ustuwani*, which was tne name given by al-Bīrunī[3] to an astrolabe on a cylindrical (orthographic) projection he devised. The *Rām Yantra* is a cylindrical instrument open at the top and having at its centre a pillar. The floor

[1] The method is very crude and the observations must have been very rough approximations only.

[2] The Jai Prakāś was known to the Arabs as al-Mastarah. For descriptions see L.A. Sédillot's *Mémoire* and Blagrave's *Art of Dyalling* quoted below (p. 86).

[3] *The Chronology of Nations* p. 357 f.

and the inside of the circular wall are graduated in scales of tangents for altitude and azimuth observations. The height of the wall from the graduated floor is equal to the distance from the circumference of the central pillar to the inside of the wall. To facilitate observation the floor is broken up into sectors (see figure 47 and plate XVII) and, consequently, as in the case of the Jai Prakāś, complementary buildings had to be constructed (see figure 41). The walls also are broken up and one section of the wall corresponds to one sector. At Delhi there are 30 sectors each of 6 degrees in each building, but at Jaipur there are 12 sectors only, and their angle is 12 degrees in one instrument and 18 degrees in the other, the spaces between them being, respectively, 18 and 12 degrees.

On each side of the wall sections are notches in which sighting bars can be placed horizontally. The construction is illustrated in plates XIII, XVII, and XXI and in figures 47, 48 and 49. Figure 41 gives a good view of the buildings as a whole. Examples of the *Rām Yantra* exist at Delhi and Jaipur only and the Jaipur instrument is quite a modern one.[1]

The dimensions of the two instruments are :

	Inside diameter.	Height.	Diameter of pillar.
Delhi	54′ 7½″	24′ 8″	5′ 3¼″
Jaipur	23′ 1″	11′ 4″	2″

31. The **Digaṁśa Yantra** ('Azimuth instrument') although not actually mentioned by Jai Singh in the preface is given, however, in Jagannāth's list (see p. 3). It is a simple and useful instrument and examples of it still exist at Jaipur, Ujjain and Benares. The instrument consists of a pillar surrounded by two circular walls (see plates XXI and XXVI, DD). The central pillar is generally about 4 feet high and the inner wall the same height while the outer wall is twice that height.

Cross wires are stretched from the cardinal points on the outer wall and both walls are graduated. The inner wall is a convenient height for a man to walk on and to look over the outer wall. By the aid of a movable string and an assistant, azimuth (horizontal angles) observations can be made with fair accuracy. The instrument may be described as a fixed large circular protractor.

The dimensions of the several *Digaṁśa Yantras* are—

	DIAMETERS.		HEIGHTS.	
	Outer wall.	Inner wall.	Outer wall.	Inner wall.
Jaipur	27′ 0″	17′ 6″	6′ 5″	3′ 2¼″
Ujjain	36′ 10″	24′ 4″	8′ 10″	4′ 6″
Benares	31′ 6″	21′ 0″	8′ 4″	4′ 1″

[1] It was built in 1891. There are also two small *Rām Yantras* which were constructed as models.

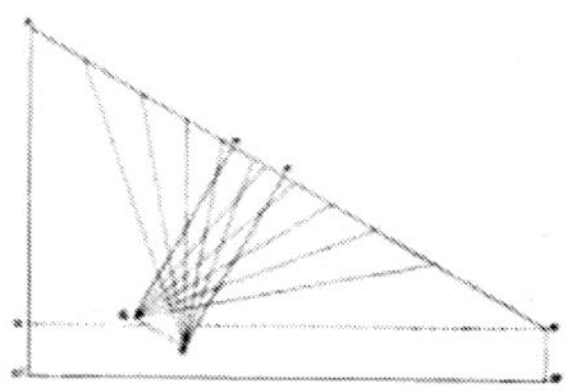

FIG. 34. DIAGRAM OF SAMBAT YANTRA.

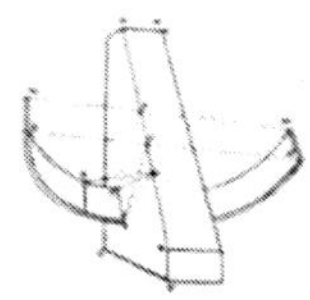

FIG. 35. DIAGRAM OF SAMBAT YANTRA.

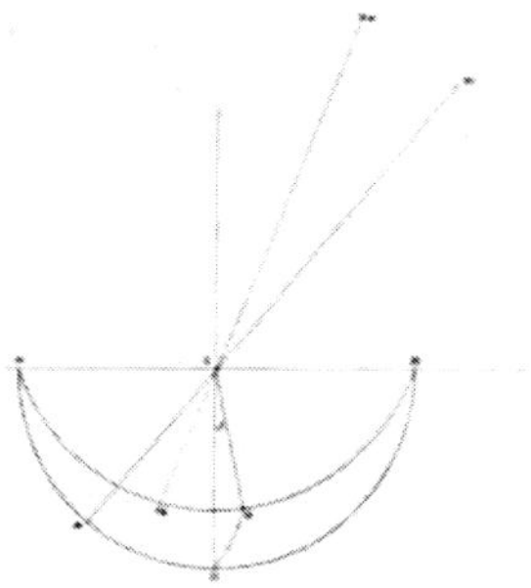

FIG. 36. CONSTRUCTION FOR THE NIYAT CHAKRA

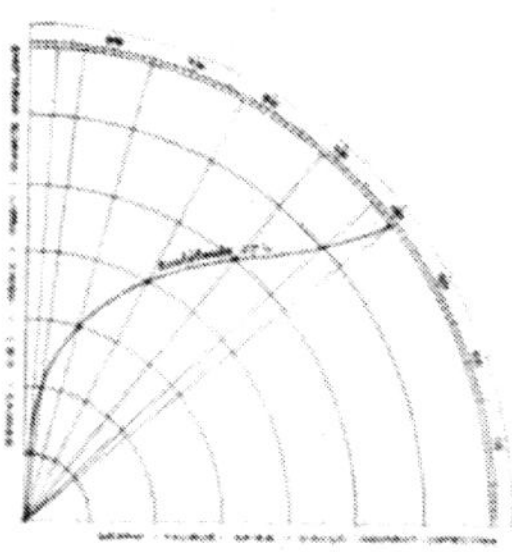

FIG. 37. DECLINATION GRAPH.

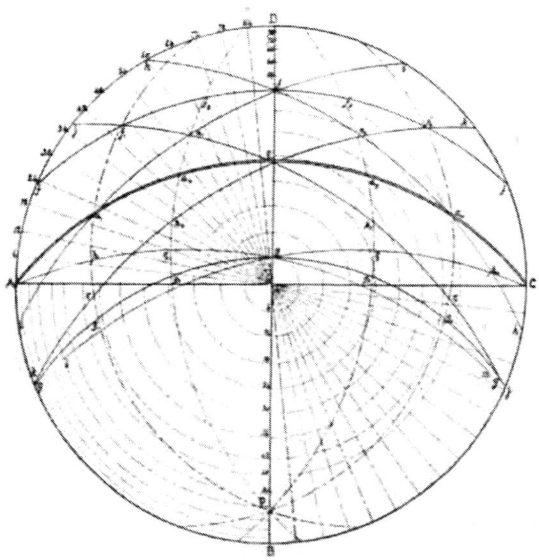

FIG. 38. DIAGRAM FOR THE JAI PRAKAS.

Jagannāth's description of the Digaṁśa Yantra is as follows :—"Make a circle on the ground with any radius. This circle is called the horizon. We shall have to make three horizons here. On the first circle build a solid pillar. On the second circle build a ring-like wall, as high as the pillar on the first circle. On the third circle make a ring-like wall twice as high as either of the former. On all these horizons mark the east to west and north to south lines, and degrees and minutes. Stretch tightly two threads across the exterior wall to represent the east and west and north and south lines, intersecting at right angles over the centre of the horizons. At the centre of the pillar, fix securely one end of a string, to the other end of the string fasten a stone and place it over the edge of the third horizon. This thread is called the 'thread of the circle of vision.'"

32. The **Narivalaya Yantra** ('Circular dial') is mentioned by Jagannāth and it occurs at Jaipur, Ujjain and Benares. It may be described as a cylindrical dial—the axis of the cylinder pointing north and south, and the northern and southern faces being parallel to the plane of the equator. At the centre of each face, and at right angles to it, is an iron style surrounded by circles, graduated into hours and minutes and *ghaṭis* and *palas*, respectively. The shadow of the style marks the time of the day, and the instrument also shows, very effectively, the passage of the sun across the equator (the equinoxes). See figures 53 and 65. Jagannāth remarks about this instrument, that it is not of much value, because it only gives readings for northerly observations. This applies to some extent to the Benares instrument (see figure 65), but not to those at Ujjain and Jaipur. The latter, however, is quite modern.

33. The **Dakshinovritti Yantra** ('Meridian circle') is like the mural quadrants found in most mediæval observatories. It consists, essentially, of a wall in the meridian, and on the wall are two graduated quadrants and centre pins (see figures 56 and 62 and plates XXIV and XXVI), which were used for observing the altitudes of heavenly bodies when passing the meridian. The instrument corresponds to the modern transit circle. Originally there was one at each observatory, but that at Delhi has been destroyed.

33(*a*). The **Shashtāṁśa Yantra** ('Sextant') occurs at Delhi and Jaipur only, and is really another form of meridian circle. It is a large graduated arc, lying in the meridian, and is built in a 'dark room' at the bottom of the masonry work, that supports the huge quadrants of the Samrāṭ Yantra. A small orifice some 30 or 40 feet above, admits the light of the sun at noon, and the image of the sun on the graduated arc marks with fair accuracy the sun's altitude. It is thus the aperture dial of the Muslims (see p. 82). At Jaipur there are two 'dark rooms,' one under each quadrant of the Samrāṭ and in each room are two arcs, the radius of each being 28 feet 4 inches. The 'dark room' at Delhi is at present inaccessible.

34. Of other masonry instruments, there are the Miśra Yantra ('mixed instrument') at Delhi and the Rāśi Valaya (zodiac dials) at Jaipur. There are some indications that these two instruments, or rather sets of instruments, were not devised by Jai Singh, and, therefore, they will be described in detail when the

observatories at Delhi and Jaipur are dealt with. The most notable feature of the Miśra Yantra is the set of arcs for meridians at Greenwich and Zurich on the west and two corresponding places on the east. The Rāśi Valaya is a set of twelve dials connected with the rising signs and which show the sun's latitude and longitude.

35. Of these instruments it is claimed that Jai Singh devised the Samrāṭ Yantra, Jai Parkāś and Rām Yantra. The evolution of these instruments will be dealt with in a concluding chapter but it may be remarked here that Jai Singh's ingenuity was chiefly concerned in the transference of designs previously executed in instruments of comparatively small size to huge masonry instruments. No new invention in the ordinary sense of the word was attempted. The Samrāṭ Yantra is in principle a very simple form of sun-dial, but it is an efficient instrument and, as Jai Singh designed it, a dignified structure. There is, so far, no evidence to show that the tangent scales on the edge of the gnomon had been previously used as on the Samrāṭ Yantra ; therefore, besides the general design, we may credit Jai Singh with this device. The Jai Prakāś or ' invention of Jai ', as it may be called, and which Jagġannāth calls the 'crest jewel' is really a sort of combined armillary sphere. Possibly the astrolabe projections suggested the idea to Jai Singh. In a manuscript copy of a work by Abdul Ali Barjendi[1] (died A.D. 1523) most of the details of such an instrument as the Jai Prakāś are given. Jai Singh's duplicated instrument is however his own design and probably the introduction of the culmination sign lines must be attributed entirely to his own invention.

The Rām Yantra is only original with respect to its size and the duplication of the instrument. The meridian lines of other places on the Miśra Yantra at Delhi was not a new idea. It had, at any rate, been worked out for vertical gnomons. The Rāśi Valaya seems to be entirely original, but it is of doubtful utility as an instrument for observation.

---

[1] Abdul 'Ali b. M. al-_osein, Nizām al-Dīn al-Barjendī wrote a commentary on Naṣīr al-Dīn's recension of the *Almagest*, a commentary on Naṣīr al-Dīn's treatise on the astrolabe, a comment on Ulugh Beg's tables, a treatise on astronomy, etc. I am indebted to Khān Bahādur Pīr Mazaffar Aḥmed of Delhi for the loan of this manuscript, which was written out by one Qabād b. 'Abdul Jalil in A. H. 1066 (=A.D. 1655) at Hyderādbād. Its date and its presence at Delhi suggests the possibility of its having once belonged to Jai Singh's library.

Fig. 39 GENERAL VIEW OF THE DELHI OBSERVATORY

Fig. 40 THE SAMRAT YANTRA, DELHI

Fig. 41 VIEW FROM THE GNOMON, DELHI

Fig. 42 VIEW FROM THE GNOMON, DELHI

PLATE XIV.

Fig. 43.

Fig. 44.

Fig. 45.

Fig. 46.

Figs. 43 & 44. DELHI OBSERVATORY AFTER DRAWINGS PUBLISHED IN A.D. [illegible]

Figs. 45 & 46. DELHI OBSERVATORY IN A.D. [illegible]

Photo-gravure Survey of India Offices, Calcutta [illegible]

CHAPTER VII.—THE DELHI OBSERVATORY OR JANTAR MANTAR.

For the Delhi observatory, known as the Jantar Mantar, we have the following approximately correct elements :—

Latitude 28° 37′ 35″ N.[1]

Longitude 77° 13′ 5″ E. of Greenwich.

Height above the sea-level 695 feet.

Magnetic declination E. 1° 45′, in 1915. Annual variation—1′.

Local time 12 minutes 12 seconds after standard time.

36. The observatory is 3 miles 3½ furlongs almost due south from the Pir Ghaib, the Trigonometrical Survey point on the Ridge near to Hindu Rao's House. It is also 1 mile 7½ furlongs 32° west of south from the Jama Masjid. In the projected new city the observatory borders (on the east) the road leading from the railway station to the Secretariat and Government House. It consequently will be a notable feature in the Imperial Capital and, apart from its historical value, it is desirable that it be made, by suitable surroundings and proper restoration, as dignified as possible.

The general plan[2] (Plate XIII) of the observatory shows the following structures :—

(*a*) The *Samrāṭ Yantra* ('Supreme instrument') a huge equinoctial dial. Figures 40, 43-46 and plate XV.

(*b*) The *Jai Prakāś* consisting of two hemispherical structures just to the south of the *Samrāṭ Yantra*. Figures 41 and 42 and plate XVIII.

(*c*) The *Rām Yantra* consisting of two circular buildings to the south of the Jai Prakāś. Figures 41, 47-49 and plate XVII.

(*d*) The *Miśra Yantra* ('mixed instrument'), north west of the *Samrāṭ Yantra*. Figures 50 and 51 and plate XIX.

(*e*) Two pillars south-west of the *Miśra Yantra*.

(*f*) A measuring platform just south of the Miśra Yantra.

37. The **Samrāṭ Yantra** is the central building of the observatory. It is the largest and most imposing, although a considerable portion of it is below the surface of the earth. It is, indeed, built into a quadrangular excavation some 15 feet deep, 125 feet from east to west and 120 feet from north to south. The structure is 68 feet high, of which 60·3 feet is above the earth's surface ; 125 feet from east to west and 113·5 feet from north to south. The details are exhibited in the plans and photographs (plate XV and figures 40, etc.). The essential parts are the inclined edges of the huge gnomon and the quadrants attached to it.[3] The edges of the gnomon point to the celestial

---

[1] In 1734 Father Boudier, who helped Jai Singh, obtained 28° 37′ N. and 75° 0′ East of Paris for the Delhi observatory. Rennel quotes Boudier as giving longitude 77° 40′. (The longitude of the Paris observatory is 2° 20′ 13·5″ E. of Greenwich.) Hunter gives Lat. 28° 37′ 36″ and long. 77° 2′ 27″ E. For the latitude Jai Singh obtained 28° 39′ 0″. See page 129.

[2] The Delhi plans were prepared for me by the Public Works Department, under the superintendence of Mr. Glen, Executive Engineer.

[3] For the theory of the instrument see p. 36.

north pole, that is they make an angle (28°37′) with the horizon equal (approximately) to the latitude of Delhi and are parallel to the earth's axis. The quadrants (M K G E D, figure 35) are at right angles to the gnomon and therefore the circles of which they form part are parallel to the plane of the equator. These quadrants have each a radius of 49·5 feet and are graduated on each edge in hours, degrees and minutes[1], the scales on the northern edges being marked in English and those on the southern edges in Indian symbols. The edges of the gnomon are marked with scales of tangents as already explained (page 36, see figures 34 and 35). The shadow of the edge of the gnomon on the quadrants gives the local time. In figure 40 the time is about ten minutes to four in the afternoon. The sun's declination is found by observing which part of the gnomon's edge casts its shadow on one of the edges of the corresponding quadrant (see page 36).

In the mass of masonry work that supports the east quadrant is a chamber which contains the *Shashṭhāṁśa Yantra.* This is a large graduated arc 60 degrees in length built in the plane of the meridian ; and through a small orifice near the top of the quadrant the sun, as it passes the meridian, shines on the arc and indicates its meridian altitude, from which its declination can be directly deduced. The chamber was closed up when the observatory was restored in 1910.

On the top of the gnomon is a circular pillar which was probably used originally for rough azimuth observations, but which is now surmounted by a small sundial of the European type. This was probably constructed in 1910 : the pillar, but not the dial, appears in the Daniells' drawings (figures 43 and 44).[2]

The lower part of the structure is now, more or less permanently it seems, below the water level of the locality. The height of the water varies[3] but for a great part of the year it covers the lower part of the quadrants and the steps and prevents access to the west quadrant altogether ; and it makes the structure useless for astronomical purposes. If the instrument is to be saved, means must be taken to prevent the water percolating to the foundations.

According to Jai Singh the Samrāṭ Yantra was built of stone and lime.[4] Hunter and Thorn say that the edges of the gnomon and quadrants were of white marble and von Orlich speaks of marble staircases (see page 48). The quadrants are now faced with lime but the time graduations are well marked with a soft black stone neatly inlaid into the face of the quadrant. The graduations on the edges of the gnomon are scratched into the lime plaster surface and are becoming obliterated.

---

[*] Jai Singh gives 18 cubits as the radius and one minute as equal to a barleycorn and-a-half. See page 13.

[1] The time scales are on the upper surface of the quadrants and the degree scales on the extreme edges. The European symbols were probably introduced at the restoration in 1910.

[2] In December last a Trigonometrical Survey party was using the top of the gnomon as a point of observation and found the dial in the way. The pillar as it was originally designed was exactly suitable for their purpose.

[3] In January 1916 its maximum depth was about 4 feet while at the end of August it was 3 feet deeper.

[4] See Preface to the *Zīj Muhammad Shāhī* p. 13 above.

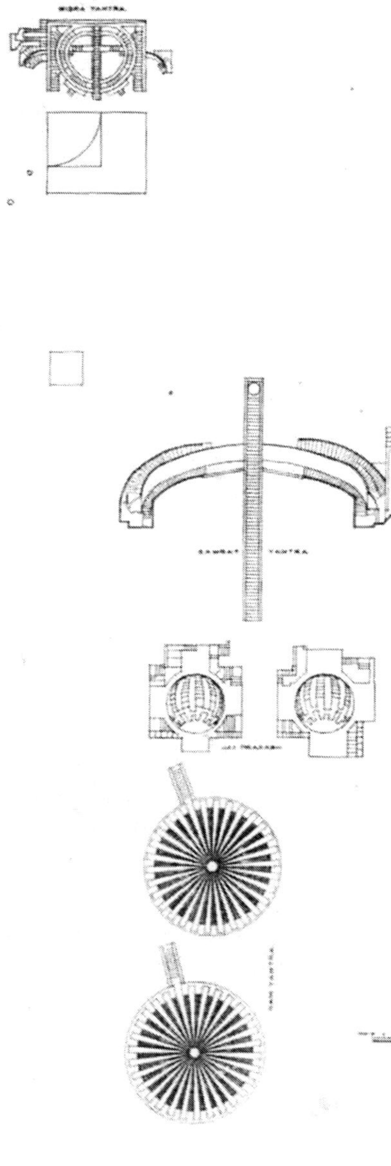

DELHI
OBSERVATORY.

SCALE

Latitude 28° 37′ 35″ N. Longitude 77° 13′ 5″ E.

Reg. No. 4388 R., 17—II. 302.

An examination of the Daniells' drawings and recent photographs (see figures 43-46), shows that only minor alterations have been made during the last hundred years. There is a slight difference in the entrance to the gnomon steps; in the old drawings is shown a set of subsidiary steps to the right of the main steps on the gnomon; and there was formerly no dial at the top of the gnomon.

38. The **Jai Prakāś** consists of two complementary concave hemispheres situated immediately south of the Samrāṭ Yantra. Their structure is best seen in plates XVIII and figures 41 and 42. Theoretically only a single hemisphere is necessary, but to facilitate observation pathways are cut into the surface; and the second Jai Prakāś has the pathways in such a way that the two instruments together show the complete surface. Cross wires were originally stretched across the hemispheres north to south and east to west, and the shadow of the intersection of these wires on the concave surface of the hemisphere indicated the position of the sun. The surface of the hemisphere is marked with altitude and azimuth circles, the tropics and intermediate circles (declination parallels), etc., so that the position of the sun can be directly read off. Also there are 'circles of the signs of the zodiac' by which the particular sign on the meridian is indicated by the position of the sun's shadow.[1] In the Delhi instruments the cross wires have been discarded although the pins to which they should be fastened are still there; and iron rods (2 inch galvanized piping) have been fixed at the centre of each **Jai Prakāś**. The pipes should be removed and the cross-wires replaced.

The descriptions given by Hunter and Thorn seem to indicate that there was a century ago only one Jai Prakāś. Hunter's words are: "Between these two buildings, (*i.e.*, the Rām Yantra) and the great equatorial dial is an instrument called *Shamlah*. It is a concave hemispherical surface, formed of mason work, to represent the interior hemisphere of the heavens. It is divided by six ribs of solid work and as many hollow places; the edges of which represent meridians at the distance of fifteen degrees from one another. The diameter of the hemisphere is twenty-seven feet five inches." Thorn uses the same phraseology. The old drawings and photographs are ambiguous on this point but they show that the original structure has been altered considerably. Probably there were two complementary instruments originally but one of them had disappeared.

39. The **Rām Yantra** consists of two large circular buildings complementary to each other situated south of the Jai Prakāś. Their general structure is best seen in figures 39, 41, 47-49 and in the plates XIII and XVII. Each consists of a circular wall and a pillar at the centre. The height of the walls and pillar from the graduated floor is equal to the inside radius of the building measured from the circumference of the pillar to the wall, *viz.* 24 feet 6½ inches and the diameter of the pillar is 5 feet 3½ inches. The walls and floor are graduated for reading horizontal (azimuth) and vertical (altitude) angles. To

[1] For a more detailed account of the theory of the instrument see p. 37.

facilitate observation the floor is cut up into thirty sectors with the spaces between of the same angular dimensions as the sectors, *viz.*, six degrees. The graduated sectors are supported on pillars three feet high so that the observer can 'place his eye' at any point on the scale. The graduated walls are similarly broken up by openings, at the sides of each of which are notches for placing sighting bars. At Delhi there are no such bars in evidence but at Jaipur they are faced with brass and carefully graduated. At Jaipur the central

Fig. 47. Rām Yantra Interior.

pillar is replaced by an iron rod. At Delhi the pillar is graduated by vertical stripes (see figure 48), each six degrees in width, and these are necessary as a point on the top of the edge (not the centre) of the pillar is the centre for which the altitude graduations on the corresponding sector and portion of the wall are made. The old descriptions and drawings show that no important structural alterations have been made during the last century. The Daniells' picture (figure 44), however, apparently shows a different entrance to the north Rām Yantra.

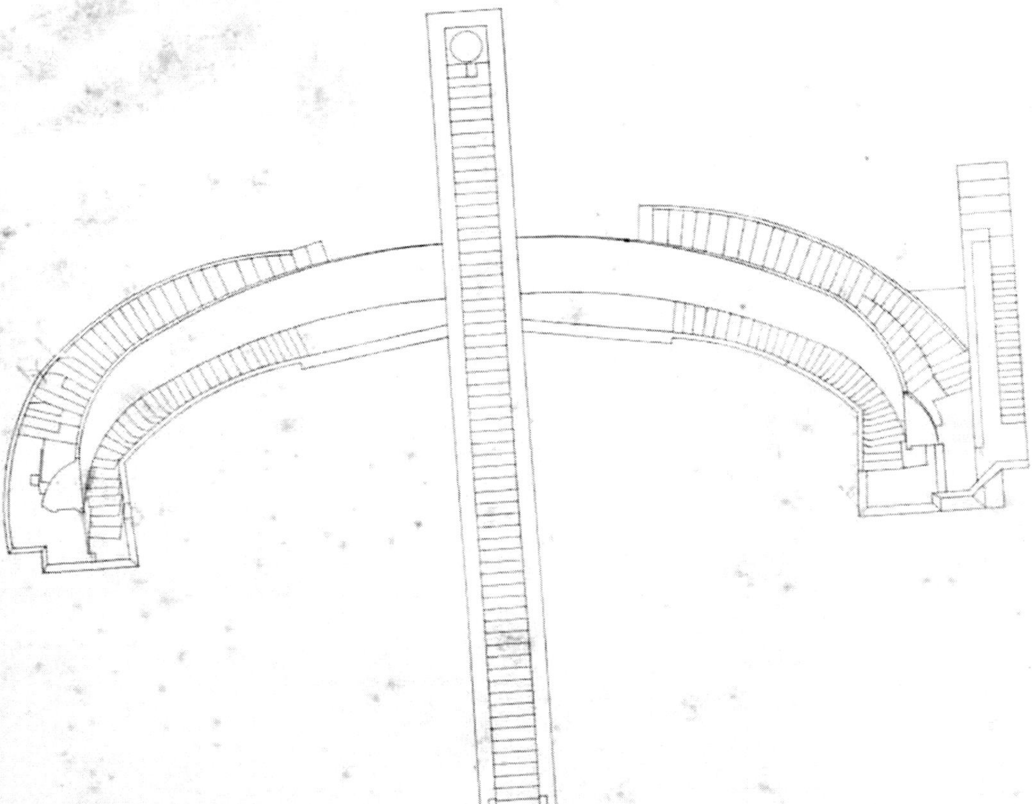

SAMRĀT YANTRA. DELHI.

PLATE XVa.

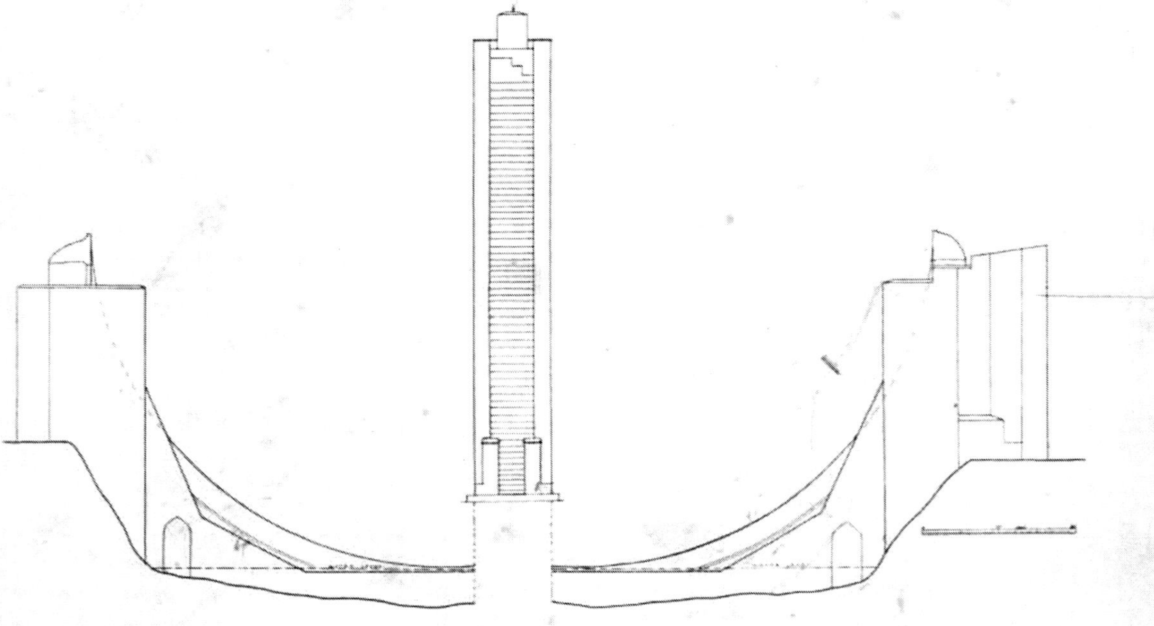

SOUTH ELEVATION

SAMRAT YANTRA, DELHI.

40. To the north-west of the Samrāṭ Yantra, and some 140 feet away is the **Miśra Yantra**, or 'mixed instrument', so named because it combines in one building, four separate instruments. Of these the *Niyat Chakra* occupies the middle of the building, and consists of a gnomon with two graduated semicircles on either side (figure 50). These semicircles lie in planes inclined to the plane of the Delhi meridian at angles of 77° 16′ W., 68° 34′ W., 68° 1′ E. and 75° 54′ E.[1]

The semicircles may be said to correspond to meridians at places whose longitudes differ from Delhi by these angles, and tradition names Greenwich observatory[2] and the observatory[3] at Zurich, "Notkey a village in Japan where there is an observatory, latitude 43° 33′ N. and Longitude 145′ 17″ E. of Greenwich" and "Serichew a town in the Pic Island in the Pacific Ocean east of Russia latitude 48° 6′ and longitude 153° 12′ E."

Let *AB* (figure 36) be the edge of the Delhi gnomon and *ABD* in the plane of the Delhi meridian. Let *ABE* make an angle δ with *ABD* then *ABE* represents a meridian at a place whose longitude difference from Delhi is δ. Let *cx*, denote the direction of the sun when it is in the Delhi meridian then the arc *DP* will measure its declination; and if CX_2 be the direction of the sun when in the plane *ABE*, then *EQ* will measure its declination. On the Niyat Yantra, the semicircle *ABD* is not marked, but *ABE* corresponds to one of the masonry semicircles, each of which is graduated north and south from E for the purpose of observing declinations.

On either side of the Niyat Yantra and joined to it is half of an equinoctial dial constructed on the same principle as the large Samrāṭ Yantra. On the west side of the building is a second quadrant, the face of which is horizontal, instead of being parallel to the axis. It is called the *Agrā Yantra* or 'amplitude instrument' and its use does not seem to have been understood by the restorers. Hunter makes no mention of this.

On the east wall of the building is a graduated semicircle called **Dakshino vritti Yantra**, used for obtaining meridian altitudes. The north wall of the Miśra Yantra is inclined to the vertical at an angle of 5 degrees (figure 51) and is marked with a large graduated circle. This is called the **Karka Rāśi Valaya** or 'Circle of the sign of Cancer.' As the latitude of Delhi observatory is 28° 37′ 35″, and the obliquity of the ecliptic is 23° 27′ 5″ nearly, the zenith distance of the sun when in Cancer is 5° 10½′ approximately, and the sun then shines over the north wall for a short period, and the shadow of the centre pin falls on the graduated circle. This may be the northern dial referred to by Jagannāth (see page 39).

---

[1] These are the angles given by the Paṇḍits but according to the measurements of the engineers who prepared the plans the angles are 77° 18′, 69° 50′, 69° 42′ and 77° 22′. They are difficult to measure accurately.

[2] This implies that the longitude of Delhi was taken as 77° 16′ E. of Greenwich. It is really 77° 31′ 6″. Zurich observatory is 8° 34′ E. of Greenwich.

[3] It may be noted that Greenwich observatory was founded in 1675 some 50 years before that at Delhi was built but that Zurich observatory did not come into existence until 1759, some sixteen years after Jai Singh's death.

41. In the front of the Miśra Yantra is a platform 47 feet by 43 feet on which are traces of a quadrant of 20 feet radius. This platform was probably used for making measurements when the instruments were being constructed or repaired.

To the south-west of the Miśra Yantra are two pillars 17 feet apart and this line joining their centres points 35° E. of north. These are mentioned in none of the accounts of the observatory. If they were part of the original observatory they probably supported one of Jai Singh's instruments such as are now found at Jaipur (see figures 28 and 29).

Hunter states that to the west of the Miśra Yantra and close to it was a wall in the meridian with double quadrants. Jagganāth, Jai Singh's assistant, recorded[1] that in the year 1651[2] of the Śālivāhana era " with this instrument the latitude of Indraprastha[3] was found to be 28° 39′ north and the maximum declination 23° 28′."

To the west of the Samrāṭ Yantra is a small building (a chowkidar's house) on which is fixed the Jaipur flag. There is a tree south-east of the eastern Jai Prakāś that partially overshadows that instrument. The tree should, of course, be removed. The whole observatory is enclosed by a mud wall about six feet high with an entrance on the west side.

42. History. The observatory at Delhi was the first one built by Jai Singh and it is here that the principal observations were made which were to form the basis of his new tables, the *Zīj Muhammad Shāhī*. There is some uncertainty about the date of construction. Paṇḍit Gokal Chand gives A.D. 1710 and Sa'id Aḥmad Khān gives 1724. The latter states that the observatory was built " in accordance with the orders of the Emperor Muhammad Shāh in the seventh year of his reign,[4] corresponding to the year 1137 of the Hegira." (=A.D. 1724-5).[5]

Jai Singh tells us[6] that he himself represented the question of preparing new tables to the Emperor who encouraged him to proceed. " To accomplish the exalted command he had received, he (Jai Singh) bound the girdle of resolution about the loins of his soul and built here (at Delhi) several of the instruments of an observatory." This seems to indicate that the construction was started *after* Muhammad Shāh ascended the throne. Also Jai Singh himself tells us that seven years were spent in preparing the tables. In 1719 Jai Singh was appointed the Emperor's lieutenant at Agra. Jagannāth records observations made at Delhi in A.D. 1729. The facts seem to point to 1724 as about the date of the founding of the Delhi Observatory.

Jai Singh tells us that at first he constructed at Delhi brass instruments of the astrolabe type in accordance with the Muslim books.[7] These he found

[1] See page 3 and Garret, page 36.
[2] A. D. 1729.
[3] Delhi.
[4] Muhammad Shāh ascended the throne in 1719 (October 9th).
[5] Thorn says : " The third year of the reign of Mohammad Shāh or 1724.
[6] See page 11.
[7] See page 12.

PLATE XVI.

Fig. 47. THE RAM YANTRA, DELHI, NORTH BUILDING.

Fig. 49. THE RAM YANTRA, DELHI, SOUTH BUILDING.

Fig. 50. MISRA YANTRA, DELHI, FROM THE SOUTH.

Fig. 52. MISRA YANTRA, DELHI, FROM THE NORTH.

Photogravure. Survey of India Offices, Calcutta, 1917.

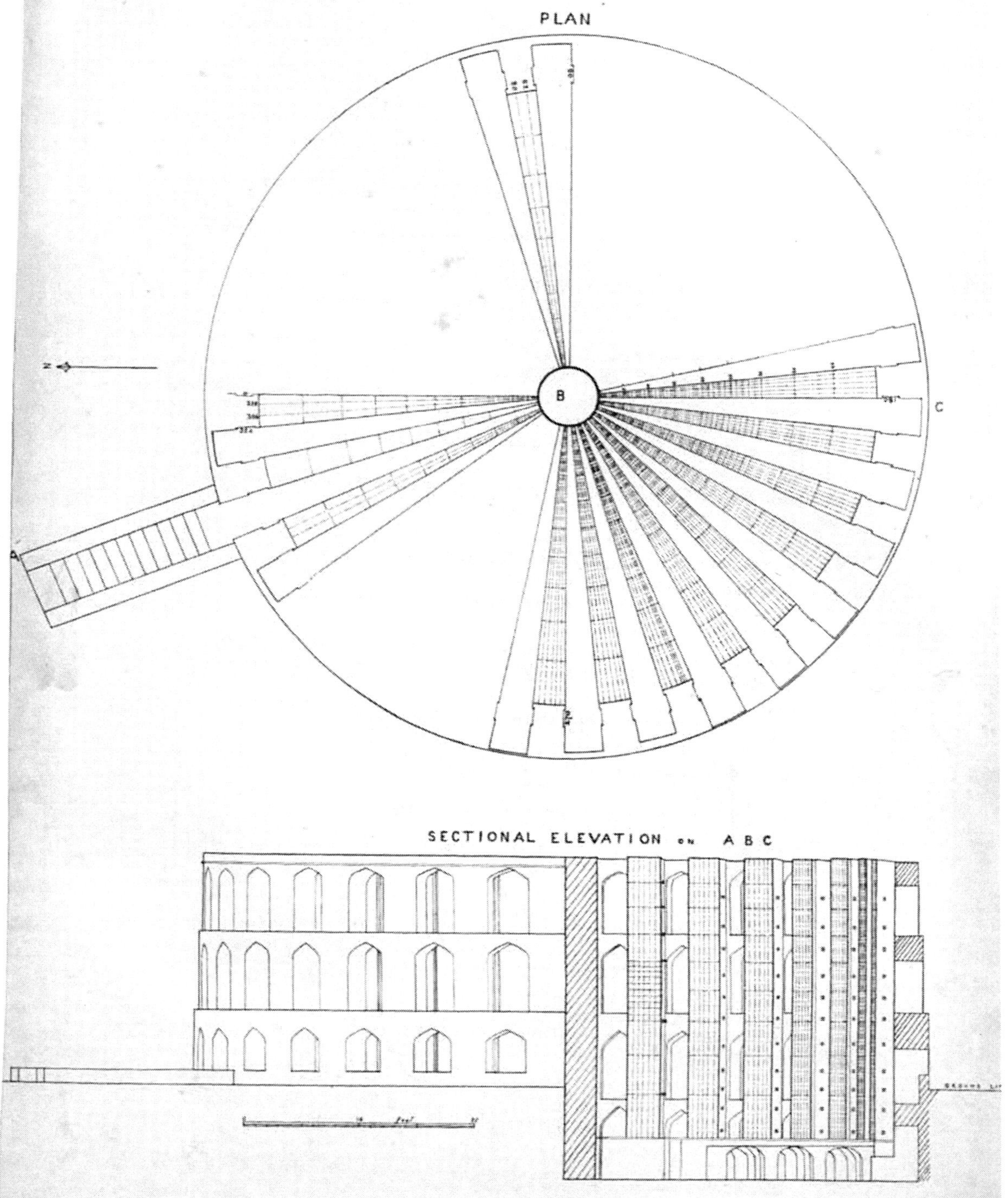

THE RAM YANTRA, DELHI.

to be unsatisfactory, and therefore he constructed "instruments of his own invention, such as Jai Prakāś and Rām Yantra and Samrāṭ Yantra .... of stone and lime of perfect stability, etc." In Jai Singh's time, therefore, the observatory probably consisted of the Samrāṭ Yantra, the Jai Prakāś the Rām Yantra, a mural quadrant, and some metal instruments. Of the present buildings possibly the *Miśra* Yantra was added by Madhu Singh "who inherited no small portion of his father's love of science."[1]

43. Early descriptions. There are fairly numerous references to the Delhi observatory in the accounts of travellers of the eighteenth and early part of the ninteenth century and some of these are worth recording. Father Claude Boudier and another priest passed through Delhi in 1734 on their journey to Jaipur (see page 6), and took observations of latitude and longitude at the observatory at Delhi. Unfortunately they have left on record no description of the observatory or the instruments.

In 1795 Franklin, in his description of the city of Delhi[2], wrote of the observatory: "It was built in the third year of the reign of Muhammad Shāh, by the Rajah Jeysing, who was assisted by many persons celebrated for their science of astronomy from Persia, India and Europe; but died before the work was completed, and it has since been plundered and almost destroyed by the Jeits under *Juhwaher Singh*."

In 1799 W. Hunter published[3] a fairly complete account of the Delhi observatory. The list of buildings and the descriptions he gives show that, to the west of the Miśra Yantra and close to it, was wall in the plane of the meridian, on which was described "a double quadrant having for centres the two upper corners of the walls . . . . One degree on these quadrants measured 2·833 inches."[4] Also, in describing the Miśra Yantra, he makes no mention of the third quadrant (Agra Yantra) on the west side. Referring to the Samrāṭ Yantra he states "It is built of stone, but the edges of the gnomon and arches, where the graduation was, were of white marble, a few small portions of which only remain."

In 1803 Major William Thorn visited Delhi and later gave a description[5] of the observatory. His description, however, is simply a summary of Hunter's and he gives no additional information whatever, although he is sometimes quoted as an authority.

Soon afterwards, the Daniells gave two illustrations[6] of the chief features of the observatory. These are here reproduced (figures 43 and 44) and they show that during the last hundred years very little alteration has really taken place; but they show some small differences, which have already been mentioned.

---

[1] Tod ii, 372.

[2] *An Account of the present State of Delhi.* By Lieut. Franklin. *Asiatic Reserches,* vol. iv., 1895, p. 431. Muhammad Shāh's reign commenced in 1719, and Jai Singh died in 1743.

[3] *Asiatic Researches,* v, 1799, 177*f*.

[4] He does not mean that he measured correctly to a thousandth of an inch but that it was approximately $2\frac{5}{6}$ inches. The radius was consequently about $13\frac{1}{2}$ feet.

[5] *Memoir of the War in India conducted by General Lord Lake in 1818, p. 171.*

[6] *Oriental Scenery,* 1815, part v, plates XIX and XX. The original drawings for these plates must have been made about A.D. 1794.

In 1843 von Orlich visited Delhi and made the following notes about the observatory: "It lies in the midst of many ruins; but it was never completed and has been, unhappily, so wantonly dilapidated by the Juts that the shattered ruins alone are to be seen. However, enough remains to show the plan of this fine building; the colossal sun-dials and quadrants, which rest upon large arches, are formed of red sandstone and bricks and the ascent to them is by handsome winding marble stair cases."[1]

Next comes Sa'id Aḥmad Khān's description[2] which was translated by Garçin de Tassy.[3] This account is not very reliable but the original work contains some rough but valuable drawings of the instruments. We read: "Now this observatory has fallen into ruin; all the instruments are broken and all traces of the lines of division have disappeared, etc."

Later writers on Delhi give brief notices of the observatory with, in two cases[4], interesting photographs.

44. **Past Restorations.** Sa'id Aḥmad Khān tells us that in 1852 the Raja of Jaipur partially restored the Samrāṭ Yantra at the request of the Archæological Society of Delhi; and in the *Proceedings of the Delhi Archæological Society* of the 6th January 1853 we read: "It having been stated that the large gnomon of the Junter Munter had been repaired at a cost of Co.'s Rs. 442-1-10, leaving a balance of Co.'s Rs. 157-14-2 of the sum presented to the society by the Rajah of Jeypore, for the repairs of that Observatory, and this being much too small a sum to enable the Society to complete the repairs or even to build around a compound wall, which is absolutely necessary, for the security of the remains from further dilapidation, it was unanimously resolved that the Agent to the Lieutenant-Governor, Delhi, be requested to make known to the Rajah of Jeypore, through the proper authorities, the inability of the Society to complete the contemplated work, without further funds." For many years nothing further was done. In 1910, His Highness the present Mahārāja of Jaipur sanctioned the restoration of the observatory at Delhi and the work was completed in 1912. Paṇḍit Gokal Chand was placed in charge of the astronomical part of the restoration which was carefully carried out. The work comprised the restoration of the buildings, the regraduation of most of the scales and, in the case of the Jai Prakāś, practically the reconstruction of the whole instrument. Most of the facings and the graduations were done in lime plaster but the main graduations on the quadrants of the Samrāṭ Yantra are in a soft black stone very neatly inlaid in the surface of the quadrants. The graduations in lime are already becoming obliterated. On the top of the gnomon of the Samrāṭ Yantra a sun-dial of European type was erected.

On each instrument a tablet giving the name of the instrument, the date of

---

[1] *Travels in India*, London, 1845, p. 49. (The account is not reliable and I am inclined to think von Orlich never visited the observatory; but what he says is the sort of thing that occurs in many guide books).

[2] *Athār al-Sanādīd* 1852. Reprinted 1876.

[3] *Journal Asiatique* V. xv, 1860, 536*f*.

[4] Carr Stephen.—*The Archæological and Monumental Remains of Delhi*, 1876; and H. C. Fanshawe, *Delhi*, *Past and Present*, 1902.

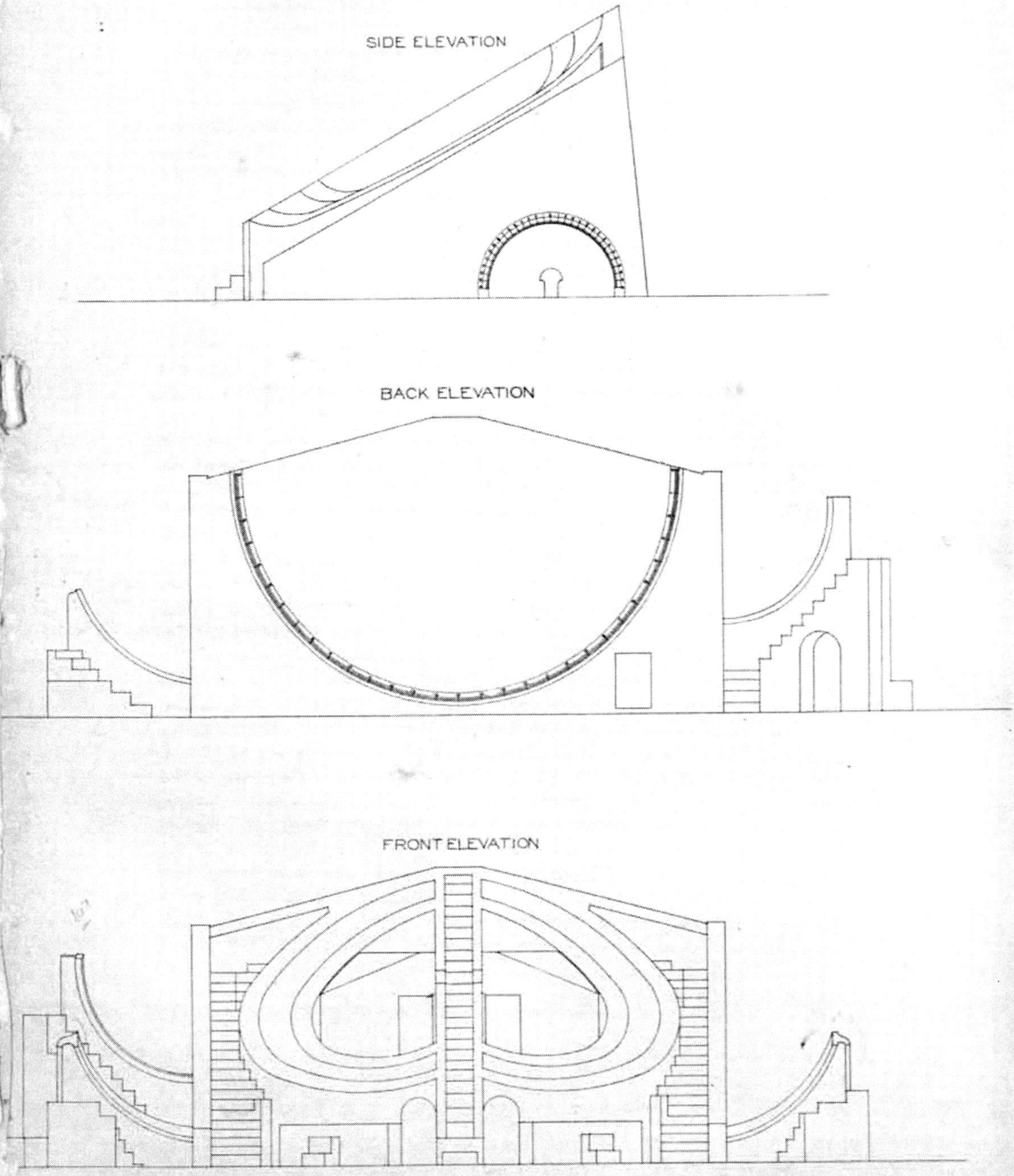

THE MISRA YANTRA, DELHI.

restoration, etc., was placed. Some of these were done in marble and some in plaster. Several of the latter are already destroyed.

45. **Future Restorations.** The Delhi Observatory buildings are worthy of permanent preservation, not only on account of their scientific and historic value, but as monuments to one of the most brilliant and remarkable princes of India, and as forming a dignified feature of the new Imperial City.

The grounds surrounding the buildings should be enclosed by a low wall and the grounds themselves should be ornamented only by a grass lawn. There should be neither trees nor shrubs, but there would be no objection to flower beds at a sufficient distance from the buildings. The buildings should be put into good order without interfering in any way with their present structure. The present pink colouring should be removed and a natural lime plaster tint substituted. The graduations should in all cases be made in some more permanent substance than lime plaster. At Jaipur marble and sandstone are both used, and at Benares the latter only. Marble, or some other suitable stone should be employed.

The Samrāṭ Yantra (figure 40, etc.) is the most important of the instruments and every effort should be made to preserve it permanently. Its foundations are in a rectangular excavation, which is now partially filled with water. Apparently the bottom of this excavation is lower than the surrounding water level, and consequently the water percolates, and covers the lower portions of the instrument. Not only is it damaging the structure, but it makes it useless for purposes of present observation. To get rid of the water is a problem for the engineers and possibly they will decide to 'concrete' the whole of the lower part of the excavation, and install a small electric pump. Unless some such means to exclude the water are taken, the chief instrument of the observatory will be utterly ruined. When the water has been excluded the chamber containing the Shashthāṁśa Yantra described above (page 42) should be opened out and the instrument put in working order.

The main graduations on the quadrants of the Samrāṭ Yantra are suitable[1] and need not be restored at present but the graduation on the edges of the gnomon need restoration badly. This necessitates the edges being refaced with marble or some suitable stone. The small dial on top of the pillar that is at the top of the gnomon should be removed. It is of no use where it now is and it prevents the pillar on which it is placed being used for its legitimate purpose. The sun-dial might be placed somewhere out of the way in the grounds. The space round the pillar is hardly sufficient for working purposes and it would perhaps be as well to reduce the diameter of the pillar or to place around it a railing for protection. (An examination of the Jaipur gnomon shows that some such arrangement would not be in opposition to Jai Singh's idea.) The top of the pillar should be graduated, as most probably it was originally, for rough azimuth observations, and should be made perfectly level.

[1] It was a mistake, I think, to introduce European measures and symbols and I should like to see the edges faced with marble as originally they were, and the old graduations replaced.

The position is one for observation and could even now be used, in the spirit of the original design, for many purposes.

The graduated parts of the Jai Prakāś require refacing either with marble or other suitable stone. The original was in lime plaster, but it did not last very long; and in 1910 the facing was again done in lime plaster but the graduations are already becoming obliterated (see figures 32 and 33). The central iron rod (galvanised piping) should be removed and the cross wires replaced.

The graduation (in lime plaster) on the walls of the Rām Yantra are not so exposed, and consequently not so liable to deteriorate as those in the Jai Prakāś. The walls of the Rām Yantra at Jaipur are in marble, but there the instrument is much smaller than that at Delhi.

The Miśra Yantra graduations are all in lime plaster, and should all be done in stone or marble. This means refacing the edges of the gnomon, and the semicircular meridians, and inlaying on the quadrants, etc.

The mural quadrant described by Hunter (see page 47), and no longer in existence, might be rebuilt. It was originally to the west of the Miśra Yantra but the space is somewhat limited there. There are examples at Jaipur (figure 56), Ujjain (figure 62), and Benares to serve as models.

The probable use of the two pillars marked on the general plan has been already explained (page 46). A brass instrument such as the Unnatamsa Yantra or large Yantra Rāj at Jaipur might be replaced.

The tablets on the instruments should be restored and revised, and they should, of course, be placed where they can easily be read;[1] and should give the name of the instrument, its uses, dates of construction or restoration, the name of the original designer (in most cases Jai Singh) and the restorers. The English versions should be revised by a European astronomer.[2]

---

[1] Two of the present tablets are too distant to be read with ease.

[2] The following is an example of those now on the instruments:—

"Kark Rashi Balay Yantra, Restored A.D. 1910. Tested by Jotish Gokal Chand Bhawa, for finding the longitude of the sun when the Cancor or the point 90 in the Faliptic comes over the plane of Meridian."

Plate XX

Fig 52. GENERAL VIEW, JAIPUR OBSERVATORY

Fig 53. NARI VALAYA JAIPUR OBSERVATORY

Fig 54. RASI VALAYA, JAIPUR OBSERVATORY

Fig 55. RASI VALAYA CAPRICORNUS

CHAPTER VIII.—JAIPUR OBSERVATORY.

The elements for Jaipur Observatory are approximately as follows :—

Latitude[1]	26° 55′ 27·4″
Longitude	75° 49′ 18·7″
Height above sea level	1,582 feet.
Magnetic declination	E. 1° 45′ in 1915.
Local time	is 26 minutes 43 seconds after standard time.

46. The observatory is within the palace precincts about 200 yards east of the minaret.[1] It is in an excellent state of preservation and is one of the 'sights' of the city. The general plan (plate XXI) shows the arrangement and some details of the instruments, which are also illustrated in plates XX and XXII.

Not only are there the usual masonry instruments but also a number of brass instruments of very considerable interest: and in the Museum outside the city walls are other brass astronomical instruments that no doubt formed part of Jai Singh's astronomical equipment. These latter have already been described in some detail (p. 16 seq.): some of them are of very great interest and value. The following list enumerates all the instruments in the observatory and the more important of those in the museum :—

MASONRY INSTRUMENTS.

Samrāṭ Yantra	Plate XXI.
Shashthāṅśa Yantra	
Rāśi Valaya	Figures 54 and 55.
Jai Prakāś	Figure 30.
Kapāla	,, 31.
Rām Yantra	,, 59.
Digaṁśa Yantra	Plate XXI.
Smaller Samrāṭ Yantra	Figure 52.
Nari Valaya Yantra	,, 53.
Dakshiṇovṛitti Yantra	,, 56.

METAL INSTRUMENTS.

Chakra Yantra	Figure 57.
Krānti Yantra	,, 58.
Unnataṁśa Yantra	
Yantra Rāja or Astrolabe	Figures 28 and 29.

In the Museum.

Astrolabe A	Figures 5 and 7.
,, B	,, 6, 8, and 13.
,, D	,, 10, 11 and 14.
Zarqāli astrolabe	,, 19 and 20.
Miscellaneous	,, 26, 27.

[1] The position of the palace minaret (Isri Lāt), about 200 yards to the west of the observatory enclosure, is Lat. 26° 55′ 27·4″ N., Long. 75° 49′ 18·5″ East of Greenwich. Tieffenthaler gave for Jaipur 26° 53′ N. and 73° 43′ East of Paris. Father Boudier gave for the observatory Lat. 26° 56′ N. and Long. 73° 50′ E. of Paris. See page 6.

47. **Samrāṭ Yantra.**—The large Samrāṭ Yantra is situated at the south-east corner of the observatory enclosure. It is the largest of all of Jai Singh's instruments, being nearly 90 feet high and 147 feet long, and the radius of the quadrants, 49 feet 10 inches. It is graduated to read to seconds, but this is impossible in practice owing to the ill defined shadow (*i.e.*, due to the size of penumbra). The tangent scales on the edge of the gnomon (see p. 36) cannot now be used, owing to the instrument overlooking the palace zenana enclosure. The readings of the quadrants appear to be slightly inconsistent, the eastern quadrant giving readings that are about two minutes out as compared with the time registered on the western quadrant.

The general structure is the same as that of the Delhi instrument, but it is of somewhat more elaborate construction and on a larger scale. Like the Delhi instrument, the foundations are below the ground level, but the floor is *pukka*, and proper arrangements are made for drainage.

Shashtāṁśa Yantra.—The Shashthāṁśa Yantra, or sextant instrument, is a huge convex arc of 60 degrees and of 28 feet 4 inches radius lying in the meridian. There are two pairs of such arcs built into the masonry that supports the east and west ends of the Samrāṭ quadrants. A small hole in the roof of each structure allows the sunlight to fall on the arc at noon. The instrument is capable of giving very accurate results but the readings are said to be in error to about 4 minutes.

Rāśi Valaya Yantra.—The Rāśī Valaya, or 'ecliptic instrument' consists of a collection of dials, situated on a platform to the west of the Samrāṭ Yantra (Plate XXI and figures 54 and 55). There are twelve such dials, one for each sign of the zodiac (that for Capricornus is seen in figure 55), and each instrument is exactly of the same type as the Samrāṭ Yantra, but the quadrants lie, not in the plane of the equator, but in the plane of the ecliptic when the particular sign is on the horizon, and the edge of the gnomon then points to the pole of the ecliptic; consequently, at the proper moment the instrument indicates the sun's latitude and (with appropriate graduations) longitude. The radius of the quadrant is 5 feet 6 inches in the case of four instruments and 4 feet 1½ inches in the case of the other eight. The contemporary lists do not mention the Rāśī Valaya.

Jai Prakāś.—The Jai Prakāś is constructed in the same manner as the Delhi instrument (Plate XXI and figure 30). The principle of the instrument has already been explained (*p.* 37). It was completely restored in 1901, in white marble, and the various circles were then marked in different colours.[1] It is 17 feet 10 inches in diameter. The instrument shows time and declination and the signs on the meridian.

Kapāla.—The Kapāla is another hemispherical instrument and is peculiar to Jaipur (figure 31). There are two examples—one being a hemisphere with its upper edge representing, as in the Jai Prakāś, the plane of the horizon, while in the other it represents the solstitial colure. The latter indicates 'rising signs' instead of meridian signs. Each Kapāla has a diameter of 11⅓ feet and is a

---

[1] The colours have since disappeared.

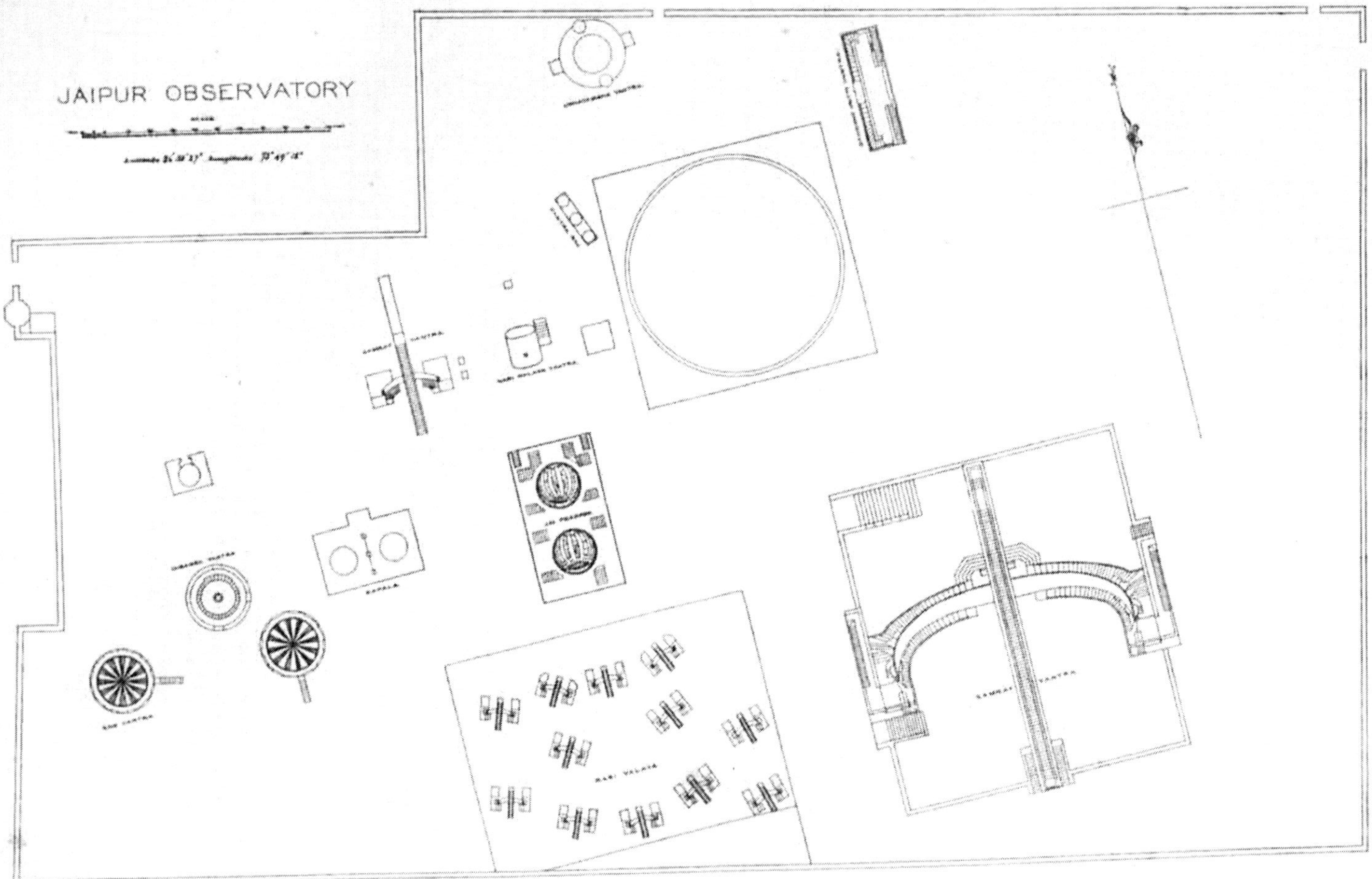

GENERAL PLAN.

complete hemisphere, that is, no pathways are cut as in the Jai Prākāś. The graduated rims are in marble, but the remainder of the surfaces are in lime plaster.

Rām Yantra.—There are four instruments in white marble (plate *XXI* and figure 59), but all of them are quite modern (? 1891), and were made according to Jagannāth's instructions. In principle they are exactly the same as the instruments at Delhi, but are much smaller, the larger pair being 23 feet 11 inches in diameter. The sectors are twelve in number occupying 12° each in one instrument and 18° in the other. For an explanation of the construction see p. 37.

Digamśa Yantra.—The Digamśa Yantra or azimuth instrument has already been described (p. 38). There is no such instrument at Delhi but there are examples at Ujjain and Benares (Plate XXI).

Nari Valaya Yantra.—There are similar instruments, but much smaller, at Ujjain and Benares. The instrument at Jaipur is a masonry cylinder some 10 feet in diameter.[1] The axis of the cylinder is horizontal and in the plane of the meridian, and the parallel faces which form the dials are in the plane of the equator. The dials are graduated into *ghaṭis* and *palas*, and also hours and minutes. According to Garrett, the southern face was added by Jai Singh's grandson, Mahārāja Purtāp Singh.

Dakshinovritti Yantra.—The construction of the Dakshiṇo Vṛitti Yantra or mural quadrant is clearly seen in plates XXI and figure 56. It is of the same principle of the similar instruments at Ujjain and Benares (that at Delhi has disappeared). On the east face are two quadrants of 20 feet radius, and on the west face is a semi-circle of 19 feet 10 inches radius. It was used for taking meridian altitudes.

The metal instruments have already been described in detail (page 16 seq.).

48. **History.**—Jaipur city was built about A. D. 1728, and the observatory was constructed about A. D. 1734. The earliest detailed description is that by Tieffenthaler, a Jesuit Missionary, who travelled in India from 1743, the date of Jai Singh's death, to 1786; but the earliest reference to the observatory is, possibly, that by Father Boudier, who, with another priest, visited Jaipur in 1834 and made observations. He makes no references to the instruments[2] and they were possibly only in the course of construction at the time of his visit. He, however, made observation, for the determination of the latitude and longitude of the observatory itself.[3]

Tieffanthaler's description[4] of the observatory is as follows:—

"But a place that deserves detailed description is that where astronomical observations are made: it is such a work as is never seen in this part of the world, and, by the novelty and grandeur of the instruments, strikes one with

[1] In the general plan (*plate XXI*) the supporting masonry work is not shown but see figure 53.

[2] Neither does he make any reference to the instruments at Delhi although he made observations there long after the observatory was built.

[3] *Lettres édifiantes*, etc., pp. 269-290. See page 6.

[4] *Des Pater Joseph Tieffenthaler's historisch—geographische Beschreibung von Hindustan*, 1785, vol. I, p. 244 *f*. French edition, p. 316 *f*.

astonishment. This large and spacious observatory is close to the King's palace. It is situated on a plain surrounded by walls and was constructed especially for the contemplation of the stars.

"On entering one first sees the twelve figures of the Zodiac all arranged in a large circle and made of purest lime. Next are seen diverse sections of the astronomical sphere, fixed according to the height of the pole at the place—with diameters of 12 or more Paris feet,[1] and besides these some large and small equinoctial dials and some astrolabes made in lime, also a meridian line and a horizontal dial engraved on a very large stone.

"But what attracts most attention is a gnomon (*axis mundi*) remarkable for its height of 70 Paris feet[2] and for its thickness—constructed in brick and lime, situated in the plane of the meridian, with an angle equal to the height of the pole. On the summit of this gnomon is a belvedere which overlooks all the town and is so high that it makes one giddy. The shadow of this gigantic gnomon falls on a prodigiously large astronomical semi-circle, of which the horns are turned towards the sky. It is artistically constructed in whitest lime or gypsum and is graduated in degrees and minutes. In the morning the shadow falls on the western quadrant and in the evening on that towards the east, and, as the gnomon lies between both the quadrants, the sun's altitude can be found at any moment. A double dial, constructed also in lime, is seen near these quadrants. It is enclosed in a kind of chamber on either side of which it is raised. When the sun passes the meridian a ray of this star enters through each of two holes pierced in a sheet of copper, and when these rays fall exactly on the middle of the two quadrants, low in summer and higher in winter, the sun is in the meridian sign and its meridian height is indicated.

"The instruments which follow have similar graduations: there are three very large astrolabes cast in copper, suspended by iron rings; a circle also of cast copper, fitted with a rule (or alhidade) and elevated at the height of the pole, suitable for determining the declination of the sun—for when you turn this instrument towards the sun it will indicate the declination on the ground.

"I pass over in silence other less important instruments, but a matter which detracts from the value of the observatory is that, in a low situation surrounded by walls, the observer cannot see the rising and setting of the stars; also the dial, gnomon and other parts being in lime plaster prevent one from making very exact observations."

49. **Restorations.**—The Jaipur observatory, being situated in the palace precincts, has been carefully preserved and has been added to from time to time. Possibly the Rāśi Valaya was added after Jai Singh's reign and possibly some of the brass instruments were brought from the Delhi observatory, but we have no direct information on these points. Some additions appear to have been made in 1891 and in 1901 His Highness the present Mahārāja decided to restore the observatory completely. The services of Lieutenant A. H. Garrett, R.E., were lent and the work was finished in 1902, in which year also Lieute-

[1] 12 Paris feet = 12·8 English feet approximately.

[2] 70 Paris feet = 74½ English feet approximately.

nant Garrett, assisted by Pandit Chandradhar Guleri, prepared and published an account of the observatory. This account, as far as the descriptions of the instruments are concerned, is an excellent one, but the parts relating to the history of astronomy are not so reliable. It is difficult to judge of the work of restoration as no reliable account of the observatory before the restoration took place is available. Colonel Hendley in 1886 gave a rough plan of the observatory and a list of the instruments[1] but not in sufficient detail for purposes of comparison.

Lieutenant Garrett's work has been somewhat severely criticised in one respect.[2] He found the Rāśi Valaya instruments did not accord altogether with his idea of their use and altered the angles slightly. He assumed that each instrument corresponded to a 'rising sign' and was to be used at the time of rising of the particular sign for the measurement of the celestial latitude and longitude. The following table shows the actual alterations made in the twelve instruments corresponding to the twelve signs.

Signs.	Measurements of gnomons before alteration.		Measurements of gnomons after alteration.	
	Azimuth.	Altitude.	Azimuth.	Altitude.
	° ′	° ′	° ′	° ′
Aries	—26 0	27 0	—25 56	24 32
Taurus	—21 30	15 30	—21 17	14 25
Gemini	—12 30	7 0	—12 19	6 36
Cancer	0 0	3 30	0 0	3 28
Leo	12 30	7 0	12 19	6 36
Virgo	21 30	15 30	21 17	14 25
Libra	26 0	27 0	25 56	24 32
Scorpio	26 0	38 0	25 37	35 33
Sagittarius	18 0	46 30	17 40	45 42
Capricornus	0 0	50 30	0 0	50 22
Aquarius	—18 0	46 30	—17 40	45 42
Pisces	—26 6	38 0	—25 37	35 33
		Maximum alteration	0 29	2 28

[1] London Indo-Colonial Exhibition in 1886. *Handbook of the Jeypore Courts*, pp. 59-62.

[2] *Indian Antiquary* XXXV, 1906, p. 34.

CHAPTER IX.—UJJAIN OBSERVATORY.

Latitude	23° 10′ 18″ N.
Longitude	75° 47′ 0″ E. of Greenwich.
Height	1500 feet.
Magnetic Declination	0° 45′ E. (1915).

Local time is 26 minutes 52 seconds after standard time.

50. The observatory is situated to the south west of the present city in the quarter called Jaisingpura, on the north bank of the river Sipra. From the Water Works it is half a mile west. The general situation is seen in the attached map and in figure 60. The river bank is corroding away and about a quarter of a mile to the east of the observatory is seen the remains of a well now standing in the river itself. The observatory is 125 feet north from the river and is hardly in danger owing to this proximity; but the drainage about the observatory is not under control. There is a small and apparently fresh nullah quite close by and the foundations of the Digaṁśa Yantra have already been partly worn away.

51. The observatory now consists of the following instruments:—

(*a*) The Samrāṭ Yantra.

(*b*) The Dakshiṇo Vṛitti Yantra.

(*c*) The Nari Valaya Yantra.

(*d*) The Digaṁśa Yantra.

These are all in a state of ruin. The foundations of the Digaṁśa Yantra have evidently moved and its walls are badly cracked. The Dakshiṇovṛitti Yantra is inclined to the perpendicular at an angle of about 5 degrees. This is possibly due to the faulty structure, for the foundations for this heavy mass of masonry seem to be inadequate. The Samrāṭ Yantra is in a dilapidated state and the styles and graduation have disappeared from the Nari Valaya.

Of the **Samrāṭ Yantra** only a skeleton remains. In the general plan (plate *XXIV*) it is shown as though complete but figure 61 shows its actual present state. It is practically of the same size as the one at Benares and the smaller one at Jaipur, namely, 22 feet high, the edge of the gnomon 47½ feet, and the radius of each quadrant 9 feet 1 inch. In 1796 or thereabout, when Hunter visited Ujjain, the quadrants were divided into ghaṭis and subdivisions. From the edges of the quadrants where they intersect the walls of the gnomon lines at right angles (*GH*, *EF* figures 34 and 35) were drawn on the gnomon and perpendicular to its edge. From the points (*H*, *F*) where these lines meet the edge of the gnomon scales of tangents were marked on the edges. All these graduations have disappeared. (For the theory of the instrument see page 36.)

The **Dakshiṇovṛitti Yantra.**—('Merdian instrument') is shown in plate XXIV and in figure 62. The masonry work is fairly intact but the graduations have disappeared. The instrument was originally something like that at Jaipur (figure 56). It consists of a wall lying in the meridian and on its east face was a double quadrant the centres of which were at points near the top corners

Fig 60. UJJAIN OBSERVATORY, DISTANT VIEW

Fig. 61 UJJAIN OBSERVATORY GENERAL VIEW

Fig. 62 UJJAIN OBSERVATORY DAKSHINO VRITTI YANTRA

Fig. 63 UJJAIN OBSERVATORY DIGAMSA YANTRA

PROOF FOR APPROVAL

of the wall and 25 feet apart. A portion of one quadrant is still visible, engraved in the lime plaster with which the wall is faced, but this is probably not the original graduation. On the ledge below the quadrants there are traces of a scale of tangents. On the west side is a flight of steps (figure 62) leading to a narrow platform at the top. At the south-west end of this platform is a small pillar, 2 feet in diameter: According to Hunter this was "graduated for observing the amplitude of the heavenly bodies at their rising and setting;" and was called *Agra Yantra* ('amplitude instrument'). The graduations have now disappeared. At the middle of the platform and on the east side is a small projection of the parapet 2 feet 4½ inches long and deep. On this, Hunter tells us, was "constructed a horizontal dial called *Puebha Yunter*." There is no sign of this dial now.

The **Nari Valaya** or 'Circular dial' is constructed on the same principle as those at Benares and Jaipur. It is situated a few feet to the south of the Samrāṭ Yantra and consists of a cylinder 7¼ feet long 3 feet 7½ inches in diameter, whose axis is fixed horizontally in the plane of the meridian, the faces of the cylinder being cut parallel to the plane of the equator. In the centre of each face and at right angles to it was an iron style round which was a circle graduated into *ghaṭis* and subdivisions. The styles and graduations have disappeared.

The **Digamśa Yantra** is similar to the one at Benares. It is situated quite close to the Samrāṭ Yantra on the east side and consists of an outer circular wall 36 feet 10 inches in diameter and 8 feet 10 inches high. Concentric with this is another wall 24 feet 4 inches in diameter and 4 feet 6 inches high. Originally there was a pillar at the centre but it has been removed. Cross wires were stretched north to south and east to west on the outer wall. At the four points of the compass in the outer and inner walls were arched openings, but all of those in the outer wall, except that to the west, have been filled up. The outer walls are badly cracked and a great part of the foundations is now exposed. This is due to the bad drainage of the slope to the river. The nullah that passes close by the Digamśa Yantra could easily be diverted. In Hunter's time the building was "roofed with tiles and converted into the abode of a Hindu deity" so that Hunter was unable to examine its construction. This is of interest as showing that even in the eighteenth century the instrument was no longer used for astronomical purposes. Hunter also writes[1]:—"Urania fled before the brazen fronted mars, and the observatory was converted into an arsenal and foundry of cannon."

52. There appear to be no records of any astronomical instruments at Ujjain earlier than those installed by Jai Singh. The date of the construction of his observatory is uncertain but it was probably between A. D. 1728 and A. D. 1734 (see page 139). There is no record as to whether or not the instruments were ever used for actual systematic observation but we know that before the end of the eighteenth century they had ceased to be so used.

[1] There is some ambiguity in Hunter's reference but Fanny Parkes (*Wanderings of a Pilgrim*, etc., II, 209) turns it into a certainty for she says: "The observatory at Oujein has since been converted into an arsenal and foundry of cannon." Her information was obtained from Hunter.

The earliest known description of the Ujjain observatory was by the Jesuit priest, Tieffenthaler, who travelled in India from 1743 to 1786. His account[1] of the observatory is as follows:—

"Not far from there is a suburb built by Djésing, King of Djépour, a ci-devant governor of this province (Mālwa). An astronomical observatory is to be seen there, with instruments, made of cement: namely two equinoctial dials, one large and one small; a gnomon (*axis mundi*) elevated according to the height of the pole at this place and set in the meridian; and on either side of this is a quadrant of a geometrical circle; also a dial made in lime and a meridian wall in stone."

The only other account of any value is that by Hunter (from which we have already quoted) who accompanied the Agra Resident's expedition to Ujjain in 1792-93. He briefly describes the instruments and he states that Jai Singh determined the latitude of Ujain to be 23° 10′ N. and Hunter himself took considerable trouble in verifying this result which he considered correct to the minutes.

53. **Restoration.**—There has been considerable discussion as to the best means of preserving Jai Singh's observatory at Ujjain; and it has been suggested that it should be restored and improved so as to be of help in the work of reforming the Hindu calendar. The instruments as they now stand are, however, far too dilapidated to be restored satisfactorily for practical purposes. They should be preserved as relics and only restored to such an extent as to show them more or less in their original state. To attempt to do more with them would be foolish. Originally the instruments were by no means the best devised by Jai Singh and they never were instruments of accuracy in the modern sense. The work of restoration should have for its end the proper preservation of the instruments in their original form: and to this end I make the following suggestions:—

(i) The drainage should be properly regulated. (It should be quite a simple matter to divert the drainage from the foundation of the instruments.) It also may be considered desirable to construct some protection on the river front. (ii) The ground surrounding the instruments should be levelled; the trees, removed, etc. (iii) For the particular instruments the following suggestions are made: (*a*) The Samrāt Yantra.—This should be restored on the same lines as the Benares instruments but Europeans graduations and symbols should not be employed. (*b*) The Dakshiṇovṛitti Yantra presents considerable difficulty because of its list. The only solution seems to be to take it down and rebuild it stone by stone on a secure foundation. (*c*) The Digaṁśa Yantra presents no great difficulty but parts of the walls will have to be taken down and rebuilt. Its permanent preservation is a matter of drainage. The cross wires should be replaced and the graduations on the walls remarked. (*d*) The Nari Valaya requires regraduation and the replacement of the styles.

Ujjain the Greenwich of India.

54. Ujjain (the οζηνη of the Greeks) or Avanti as it was often called, is mentioned in early Hindu Astronomical works as situated on the prime meridian and tradition also makes it the centre of astromomical learning in India.

[1] *Historische—geographische Beschreibung von Hindustan*, vol. i, 246.

[2] *Asiatic Researches V*, 1799, p. 194 f.

UJJAIN OBSERVATORY

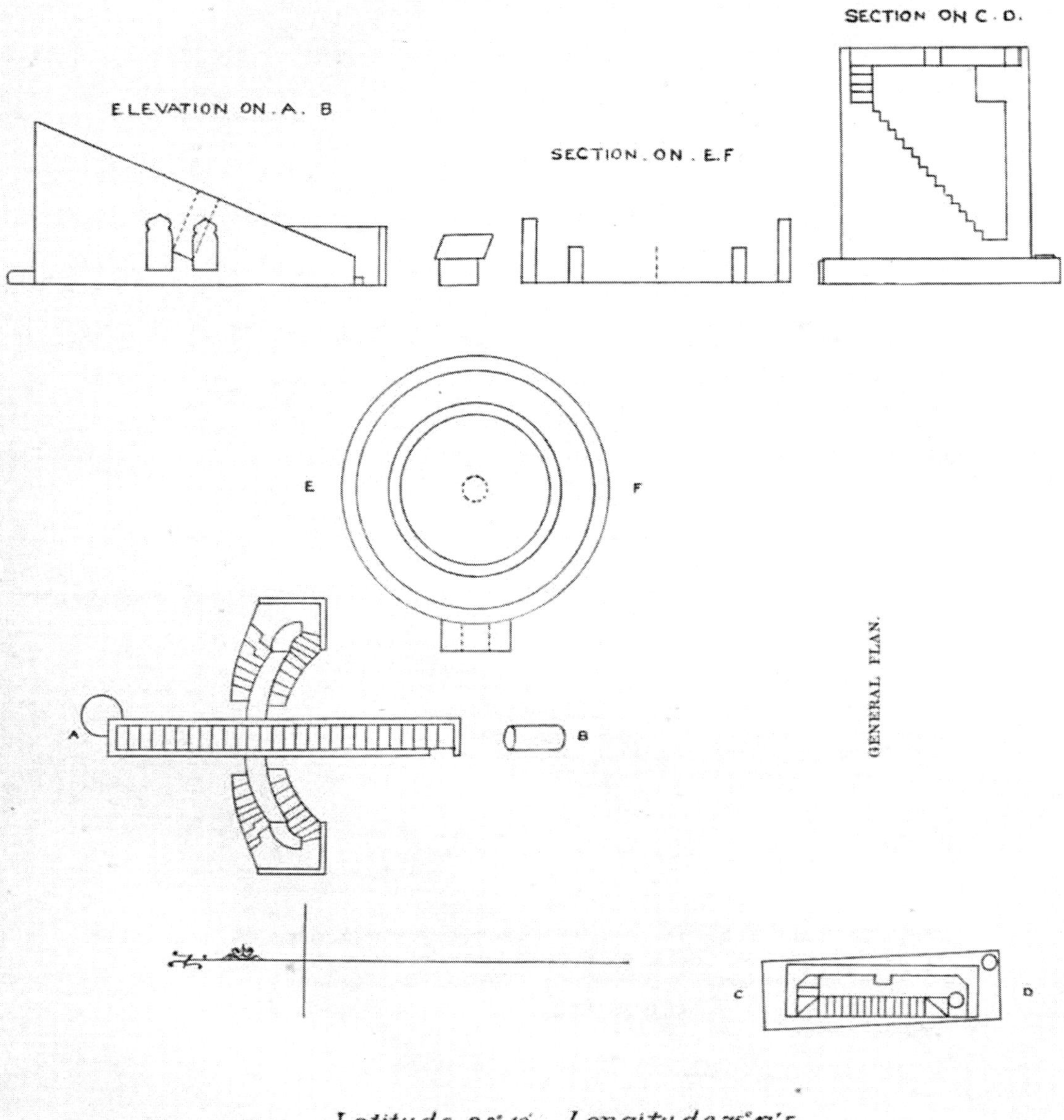

Latitude 23° 10' N Longitude 75° 47' E

0 5 0 10 20 30 40 FEET

In the *Pañchasiddhāntikā* (xiii, 17) we read: "Ujjayinī is near to Lankā, being situated to the north on the same meridian: hence the noon of the two places occurs at the same time, but their days are unequal with the exception of the equinoctial days." The latitude is given (v. 19) as 24° N. and "the *nāḍikās* arising from the difference in longitude from Yavana are seven and a third in Avantī, nine in Vārāṇasī.[1]" (iii, 13.)

The *Sūrya Siddhānta* (i, 62) says: "Situated upon the line which passes through the haunt of the demons[2] and the mountain which is the seat of the Gods[3] are Rohītaka and Avantī as also the adjacent lake."

Alberuni's discussion of the position of Ujjain is of considerable interest. He writes:[4] "All canons of the Hindus agree in this that the line connecting Laṅkā with Meru divides the οἰκουμένη lengthwise in two halves and that it passes through the city of Ujain, the fortress of Rohitaka, the river Yamunā,[5] the plain of Tāneshar, and the Cold mountains. The longitudes of the places are measured by their distance from this line. On this head I know of no difference except the following passage in the book of Āryabhaṭa of Kusumapura: —'People say that Kurukshetra, *i.e.*, the plain of Tāneshar, lies on the line which connects Lankā with Meru and passes through Ujjain. So they report on the authority of Puliśa[6]. But he was much too intelligent not to have known the subject better. The times of eclipses prove that statement to be erroneous, and Pṛithusvāmin maintains that the difference between the longitudes of Kurukshetra and Ujjain is 120 yojanas.' These are the words of Āryabhaṭa. Ya'Ḳūb Ibn Ṭārik[7] says in his book entitled the *Composition of the Spheres*, that the latitude of Ujjain is 4⅗ degrees, but he does not say whether it lies in the north or south. Besides he states it, on the authority of the book *Al-Arkand* to be 4⅗ degrees[8]. We, however, found a totally different latitude of Ujain in the same book in a calculation relating to the distance between Ujain and Almanṣūra,[9] which the author calls Brahmaṇavāṭa, *i.e.*, Bamhanwā, *viz.*, latitude of Ujain, 22° 49′; latitude of Almanṣūra, 24° 1′. According to the same book, the straight shadow in Lohāniyye,[10] *i.e.*, Loharānī, is 5⅔ digits. On the other hand, however, all the canons of the Hindus agree in this, that the latitude of Ujain is 24 degrees, and that the sun culminates over it at the time of the summer solstice."

---

[1] Yavana is Alexandria whose longitude is approximately 29° 51′ E. of Greenwich. Seven and one-third *nāḍikās* = 2 hours 56 minutes = 44 degrees and 9 *nāḍikās* = 54°. These give the longitude of Ujjain and Vārāṇasi (Benares) as 73° 51′ and 83° 51′. Their longitudes are approximately 75° 47′ and 83° 0′ 46″.

[2] Lankā.

[3] Mount Meru.

[4] *Alberuni's India* (By E. Sachau) i, 316.

[5] ? At Mathurā according to Alberuni (i, 308). Rohitaka, he says, is in the district of Multan. It was deserted in Alberuni's time.

[6] ? Fl. circa A.D. 400.

[7] ? Died A.D. 796, see Suter, p. 4.

[8] This must be an equinoctial shadow length, which gives a latitude of $\tan^{-1} \frac{4\frac{3}{5}}{12} = 20° \, 57\frac{1}{2}'$ roughly. The correct latitude is about 23° 10′ which gives a shadow of $5\frac{2}{15}$ digits.

[9] This is given on the 'Jaipur II' astrolabe with latitude 27° 40′ North. See p. 127.

[10] Perhaps the Λαγίβαρι of Ptolemy. $\tan^{-1} \frac{5\frac{2}{3}}{12} = 25° \, 15'$ nearly.

Again he says (i, 308):—"The city of Ujain, which in the tables of the longitudes of places is mentioned as Uzain, and as situated on the sea, is in reality 100 *yojana* distant from the sea, etc."

Bhāskara in his *Siddhānta Śiromaṇi* (*Gaṇita vii*, 2) writes:—"The line which passing above Lankā and Ujjayinī and touching the region of Kurukshetra and other places, goes through Meru—that line is by the wise regarded as the central meridian of the earth."

Bhāskara also mentions Ujjain in several other connexions as the place of zero longitude and he gives its latitude as 'one sixteenth of the whole circumference, north of the equator. This is equivalent to 22½ degrees north, whereas the latitude of the present city is about 23° 11′, that of Jai Singh's observatory being approximately 23° 10′ 24″. Another value was obtained from the length of the equinoctial shadow which was given as 5 *daṇḍas* 10 minutes, or 310 minutes. This[1] is the shadow of a gnomon 12 *daṇḍas* or 720 minutes high and hence the latitude $= \tan^{-1} \frac{31}{72} = 23° \ 17' \ 40''$.

55. The question of the formation of a new observatory at Ujjain is one of great importance. Ujjain is one of the most ancient astronomical centres in the world and not only should it have a modern observatory but it should be the centre of Hindu astronomical teaching. Perhaps one of the most important practical questions to settle is the position of such an observatory, which is to be the position of zero longitude for Hindu astronomers, and is to accord with the traditional position of zero longitude. To assist in this very important matter the annexed map of Ujjain has, with the assistance of the Resident of Gwalior, the Director-General of Archæology and the Surveyor-General, been produced. The Trigonometrical Survey point on the map is Hill 1678 whose longitude is approximately 75°46′44″ and latitude approximately 23° 11′ 6″ North. It is doubtful whether there ever was a fixed position in ancient Ujjain which was considered as of zero longitude. Rather vaguely the old city of Ujjain — to the north of the present city—was meant; or it is just possible that Jai Singh considered this point when he located his observatory to the south of the present city and that the site of Jai Singh's observatory is the traditional place—but this is doubtful. The plan now to follow is to fix upon the position of the new observatory and determine its longitude and latitude independently of tradition. The accompanying map should be of help in obtaining the first approximations for the longitude and latitude of such a position and it is hoped that it will be of use to the Paṇḍits of Ujjain.

[1] Guerin, *Astronomie Indienne*, Paris, 1847, p. 146.

CHAPTER X—BENARES OBSERVATORY.

Latitude	25° 18′ 24·9″ N.
Longitude	83° 0′ 46·1″ E. of Greenwich.
Height above sea level	350 feet.
Magnetic declination	E. 0° 45′ (1915).
Local time	2 minutes 3 seconds before standard time.

56. The observatory is situated on the roof of the old part of the building known as the Mānmandira, which was built by Mān Singh, a Rājah of Amber, who flourished at the beginning of the seventeenth century. (He died in A. D. 1614.) This building is on the west bank of the Ganges near the Maṇi Karṇikā ghat and 1¾ miles south-east by south from Queen's College. The proper approach is from the river front, that from the city being through narrow, unsavoury alleys. "Though not very architectural in its general appearance," writes Fergusson,[1] "(it) has on the river face a balconied window, which is a fair and pleasing specimen of his (Mān Singh's) age." (Figure 64).

On the roof of the Mānmandira as constructed by Mān Singh and a little over a century after it was built, Sawāī Jai Singh of Jaipur placed the astronomical instruments that now form the observatory. Some time about the beginning of the nineteenth century the Mānmandira appears to have been enlarged,[2] and about the middle of the nineteenth century it was restored. In figure 64 the part of the Mānmandira that supports the observatory is to the right: the circular Digaṁśa Yantra is seen above the three balconied windows.

The general plan of the roof of this part of the building (Plate *XXVI*), shows—

The larger Samrāṭ yantra (*AA*), the Narivalaya yantra, the Chakra yantra (*CC*), the Digaṁśa yantra (*DD*), and the smaller Samrāṭ yantra.

On the east wall of AA is a double quadrant or Dakshiṇovṛitti yantra and to the south of AA is another Dakshiṇovṛitti yantra not shown in the plan. The grooves shown in the plan were possibly used for levelling purposes.

57. The Samrāṭ Yantra (AA) is of the same type as those at the other observatories and is the same size as that at Ujjain and the smaller one at Jaipur. Its height is 22 feet 3½ inches, the edge of the gnomon is 39 feet 8½ inches long and the radius of each quadrant is 9 feet 1¼ inches (see p. 36). The edges of the gnomon and the quadrants are faced with sand-stone and the graduations are carefully marked. On the quadrants every half-hour is marked by two inlaid metal discs, the one towards the north edge being inscribed in Indian characters while the one on the south is in European figures. Each edge is also graduated into minutes and quarter minutes; and also into degrees and tenths of a degree. The edges of the gnomon are graduated with the usual tangent scales (see page 36). A comparison between

[1] *A History of Indian and Eastern Architecture*, vol. ii, 178.

[2] Compare Campbell's drawing (c. 1773) figure and Prinsep's (c. 1825). The latter is given in *Benares Illustrated by a series of Drawings*, by James Prinsep.

a drawing made about 140 years ago (figure 67) and a recent photograph (figure 66) shows that very little alteration has been made, the only noticeable being the inlaid metal discs already referred to, the employment of European symbols and the division into hours instead of *ghaṭis*.

On the east wall of the gnomon are two graduated quadrants (figure 66) used as a Dakshiṇovṛitti Yantra or meridian instrument. Each quadrant has a radius of 10 feet 7 inches. The shadows of one of the pins (fixed at the top of each quadrant) gives the zenith distance at noon and zenith distances of other heavenly bodies could be observed directly by moving the eye along the appropriate quadrants. Under these quadrants is a platform (shown in the plan) for the observer. Apparently in 1773 these quadrants were not in existence (see figure 67) but according to Paṇḍit Bapu Deva Śāstri they were there in 1865.

The other Dakshiṇovṛitti Yantra is a self-contained instrument consisting of a wall lying in the meridian on the east face of which are two quadrants each of 7 feet 9½ inches radius. Sir Robert Barker in 1777 stated that the quadrants were of different radii the larger of which he judged to be 20 feet. If his description be correct the instrument must have been entirely rebuilt later on, possibly when the Mānmandira was added to. In 1865 to the east of this instrument were three circles of 10 feet 3 inches, 2 feet 4 inches, and 3 feet 5 inches respectively in diameter; and also a stone square with sides 2 feet 2 inches. The circles were possibly used for construction purposes.

The smaller Samrāṭ Yantra calls for little remark. It is 8 feet 3 inches high and the radius of the quadrants is 3 feet 2 inches (figure 65). If the early drawing (figure 67) is correct the instrument has been moved from its original position.

The **Nari Valaya** ('circular dial') or *Uttara dakshiṇo Gola* (north and south dial) is a cylindrical dial—the axis of the cylinder pointing north and south, and the northern and southern faces being parallel to the plane of the equator. At the centre of each face and at right angles to it is a short iron style surrounded by two circles—the outer one (on the northern face)[1] graduated in hours, etc., and the inner one in *ghaṭis*, etc. Besides serving as an ordinary dial the instrument marks the equinoxes, since the northern face can only be used for sun observations when the sun is north of the equator. The inscription on the instrument reads:—"Narivalaya Dakshin and Uttra Gola. The use of this instrument is to find whether the heavenly bodies are in the northern or southern hemisphere. It gives time also."

The Digamśa Yantra ('Azimuth instrument') marked DD in the plan is the large circular building at the east of the terrace. It is partly visible in the general view of the Mānmandira (figure 64). The exterior diameter is 31½ feet, the outer wall is 8 feet 4 inches high and the inner wall and central pillar are each 4 feet 2 inches high, and an iron rod fixed to the central pillar is of the same height as the outer wall. The tops of both walls were originally graduated into degrees, etc.; and cross wires were stretched north to south and east to west on the outer wall. The use of the instrument is to measure azimuths

[1] There is some indication that this instrument was originally made for northern observations only. See p. 39.

or horizontal angles (see page 38), but it is now of little practical use owing to its being surrounded on all sides but one by buildings.

To the south-east of the Digaṁśa Yantra there used to be another dial whose diameter was 6 feet 2 inches. It was on a platform slightly higher than the terrace, and approached by steps. At the present time there is no space to accommodate such an instrument and Campbell's drawing of 1773 (figure 67) shows no such instrument. However, Williams mentioned that it had been excluded from Campbell's drawings[1] and it was mentioned by Hunter in 1797 and by Paṇḍit Bapu Deva Śāstri in 1865.

The Chakra Yantra is shown in the plan at CC. It is an equatorial and was common to most mediæval observatories. It consists of an iron circle (declination circle) 3 feet 7 inches in diameter, one inch thick and two broad, faced with brass on which degrees and minutes are marked. The circle is fixed so that it can revolve round an axis parallel to the earth's axis. At the southern extremity of this axis and on the pillar which supports the instrument is a graduated circle (hour circle) in the plane of the equator (figure 65). There is no pointer for this hour circle, and, according to Hunter, there was none in 1797. Attached to the centre of the declination circle is a sighter consisting of a hollow brass tube (figure 68) but this is comparatively new. Hunter wrote: "Observations with this instrument cannot have admitted of much accuracy as the index is not furnished with sights; and the pin by which it is fixed to the centre of the circle is so prominent, that the eye cannot look along the index itself." In figure 68 can be seen what is probably the old index and also the new tube sighter.

The sighting arrangement is fixed to the big circle by a pin and this pin

FIG. 69. THE FARAS OR HORSE.

[1] Phil. Trans. Royal Soc. 1793 i. 45.

is fixed by a cotter or wedge, shaped roughly into some semblance of a horse's head (figure 68). In the section on the astrolabe (page 18) it was stated that the Arabs called such a wedge *faras* ('horse') and that in mediæval Europe it was generally made into some semblance of a horse's head (see figures 24). "Thorw which Pyn," wrote Chaucer, "ther goth a litel wegge which that is cleped the hors that streyneth alle these parties to hepe." There is abundant evidence to show that the horse shaped design of the wedge was brought to India by the Muslims and the example on the *Chakra Yantra* is interesting as evidence of the persistence of a traditional design, and in some degree as evidence of the ultimate source of the design of Jai Singh's instruments.

58. **History.**—The Mānmandira was built about the beginning of the seventeenth century. Campbell's drawing of 1773 and Prinsep's of 1825 and recent photographs enable us to some extent to trace the additions and alterations made. The part of the façade that is directly under the observatory belongs to the oldest part of the building and that part that has no balconied windows is comparatively new (see figure 64). The fine window on the extreme north (right) of the building was given by Prinsep in his illustrations of Benares and has been described by Fergusson and Havell.

The astronomical instruments were added by Jai Singh about A. D. 1737.[1] The date is not certain and nearly every writer gives a different one. Sir Robert Barker, who was almost a contemporary of Jai Singh[2], said that the observatory was built by Akbar; Prinsep wrote: "The building was converted into an observatory by Jysing, A. D. 1680" and refers to a supposed description of it by Tavernier;[3] another writer gives A. D. 1693[4] and another 1700. Father Boudier who visited Benares in 1734 and made astronomical observations there makes no mention of the observatory. Jai Singh himself tells us that in order to confirm the observations made at Delhi he constructed instruments of the same kind at Jaipur, Mathurā, Benares and Ujjain; and the Delhi observatory was probably built about 1724; that at Jaipur was built in 1734 and William's date for the observatory at Benares, 1737, may be accepted.

59. **Early Descriptions.**—In A.D. 1777[5] Sir Robert Barker, who was for a short time Commander-in-Chief in Bengal, published a description of the instruments together with a perspective drawing of the observatory as a whole and detailed drawings of the Samrāṭ Yantra done by Lieutenant-Colonel Campbell, Chief Engineer of the Company's service. The perspective drawing is here reproduced (figure 67) and it shows that the main features of the observatory are the same to-day as they were nearly a century and a half ago. There are apparent differences but some of them may be due to the nature of the drawing: *e.g.*, the Narivalaya and the small Samrāṭ appear to have been

[1] This is the date given by Williams, who, on all points that can be verified, is extremely reliable. See p. 139.

[2] Sir Robert Barker lived from 1729-1789 and went out to India in 1749 six years after Jai Singh's death.

[3] Tavernier died in 1689, three years after Jai Singh's birth.

[4] This particular mistake is repeated in the *Encyclopædia Britannica*, vol. 3, p. 714.

[5] Sir Robert Barker left India in 1773. His notes and the drawings were probably made in 1872-73.

BENARES OBSERVATORY

GENERAL PLAN.

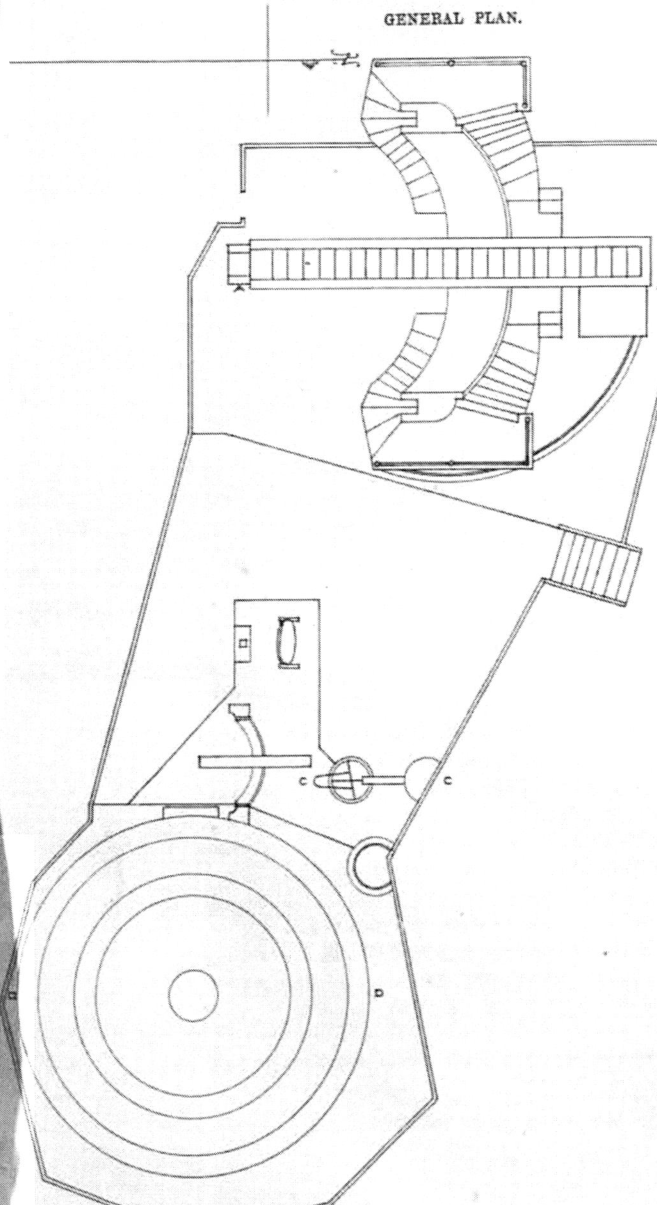

Latitude 25°18′24·9″N. Longitude 83°0′46·1″E.

10 5 0 10 20 FEET

ELEVATION ON A. A.

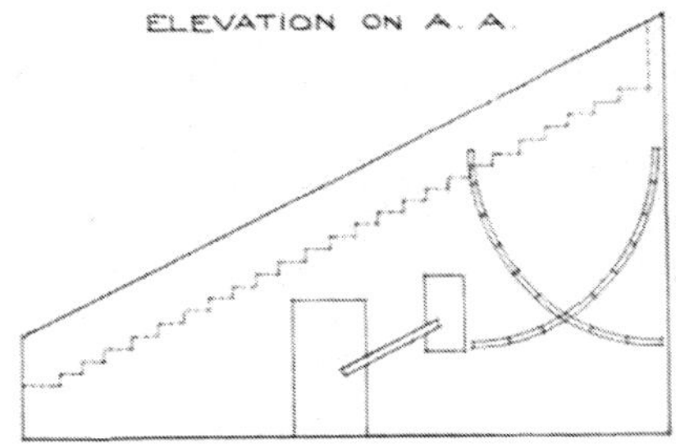

SAMRAT YANTRA

10 0 10 FEET

SECTION ON C. C.

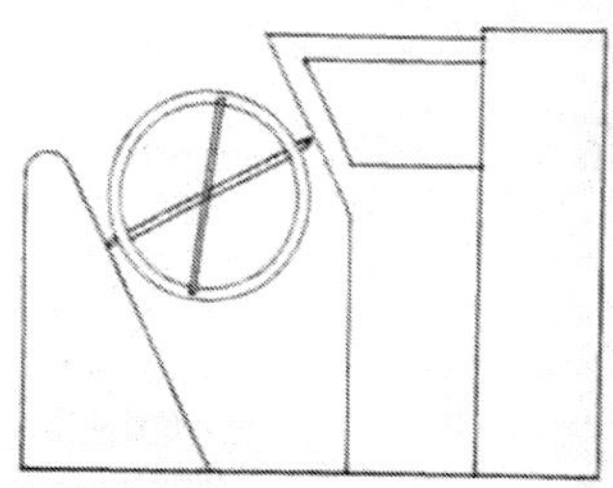

CHAKRA YANTRA.

4 3 2 1 0 4 FEET

SECTION ON D. D.

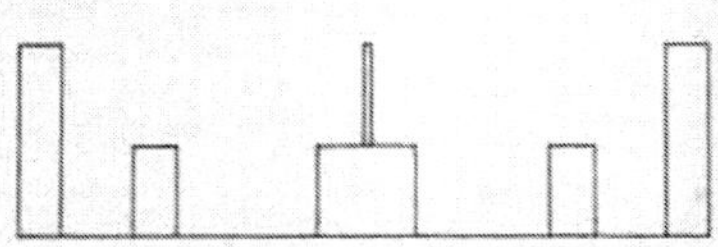

DIGAMSA YANTRA.

10 5 0 10 FEET

displaced slightly; the wall that supports the east quadrant is different in detail; there are no graduated quadrants on the east wall of the gnomon; the edge of the plinth on which the larger Samrāṭ stands has changed its alignment; and in the drawing no plinth at all is shown for the other instruments. From the notes of Williams and Hunter it appears that the drawing is somewhat incorrect, or at any rate misleading for the south east corner of the terrace, for it shows no trace of the second Nari Valaya described by them.

Sir Robert Barker's account of the Benares Observatory was of the nature of a communication to the Royal Society, London. Further information seems to have been asked for and this was supplied by Mr. J. L. Williams of Benares in A.D. 1792, who recorded very careful measurements and added some interesting details as to the history of the place. He writes: "The area, or space comprising the whole of the buildings and instruments, is called in Hindoo, *Maun-mundel*; the cells and all the lower part of the area, were built many years ago, of which there remains no chronological account, by the Rajah Mansing, for the repose of holy men, and pilgrims, who came to perform their ablutions in the Ganges, on the banks of which the building stand. On the top of this the observatory was built, by the Rajah Jetsing, for observing the stars, and other heavenly bodies; it was begun in 1794[1] Sambat, and it is said was finished in two years. The Rajah died in 1800[2] Sambat. The design was drawn by Jaggernaut and executed under the direction of Sadashu Ma Makajin; but the head workman was Mahon, the son of Mahon, a pot maker of Jepoor."

In 1799 Hunter gave a brief description of the observatory supplementary to the previous accounts. He speaks of the accuracy of Mr. William's measurements and explains some of the terms used; and clears up one or two other points. In 1848 Sir Joseph Hooker made excellent drawings of three of the instruments and in his diary records† that "the observatory is still the most interesting object in Benares although it is now dirty and ruinous, and the great stone instruments are rapidly crumbling away." The only other descriptions it is here necessary to mention are those by Paṇḍit Bapu Deva Śāstri; Lala Chiman Lal, and Paṇḍit Gokal Chand.

Restorations.—Of previous restorations we know very little. Sherring states that the Brahmans "were utterly careless" about preserving the instruments. According to Havell the Mānmandira was restored in the middle of the nineteenth century. In 1912 the present Mahārāja of Jaipur ordered the complete restoration of the instruments. This work was very thoroughly done under the direction of the court astronomer, Paṇḍit Gokal Chand.

The observatory at Benares has long since ceased to be used for practical purposes. The Brahmans consulted by Williams in 1792 all agreed that it "never was used for any nice observations." Its present situation, surrounded on most sides by buildings, is not ideal for astronomical purposes; and the instruments are of course very crude compared with those in modern observatories. The value of the observatory is chiefly historical; it is a monument to one of

[1] A.D. 1737. [2] A.D. 1743

† *Himalayan Journals*, 1854, pp. 74—77.

the brightest intellects of India; and it illustrates a very interesting phase of the history of astronomy. It might have another value if advantage were taken, namely an educational one: for the demonstration of the elements of practical astronomy a better set of instruments could hardly be devised. But, apparently, astronomy is no longer studied at Benares, Ujjain and Jaipur.

CHAPTER XI.—MATHURĀ OBSERVATORY.

61. The old fort, at Mathurā, known as *Kans ka kila* was rebuilt by Raja Mān Singh of Jaipur. On the top of this fort Jai Singh built the last of his observatories. The whole of it has now disappeared. "A little before the Mutiny," Growse† tells us, "the buildings were sold to a government contractor Joti Prasād, who destroyed them for the sake of the materials. Certainly they had ceased to be of any practical use; but they were of interest, both in the history of science and as a memorial of one of the most remarkable men, etc."

Tieffenthaler and Hunter give brief descriptions of the Mathura observatory and as these appear to be the only actual descriptions preserved they are given in full. Tieffenthaler's account (i. 143) is as follows:—

"On the roof of the fortress are seen certain astronomical instruments, erected by the famous Rajah Djésing, a lover of astronomy: principally a gnomon in lime stone which represents the axis of the earth, 12 Paris feet in height; some equinoctial dials of 5 spans in diameter; and some other smaller ones arranged for the latitude of the place; while other instruments exhibit different sections of the sphere. The observatory is only a feeble imitation of that at Djepour; but it has the advantage over the latter of an elevated situation which dominates an immense plain, while the observatory of Djepour is situated in a plain and the rising and setting of the stars cannot be seen except from the top of a masonry gnomon of prodigious height."

62. "At Matra," wrote Hunter in 1799,[1] "the remains of the observatory are in the fort, which was built by Jayasinha[2] on the bank of the Jumna. The instruments are on the roof of one of the apartments. They are all imperfect and in general of small dimensions. (1) An equinoctial dial, being a circle nine feet two inches in diameter, placed parallel to the plane of the equator and facing northwards. It is divided into degrees, which are numbered as *pals* 10, 20, 30, 40, 50, 60: lastly, each subdivision is further divided into five parts, being 12 minutes or two *pals*. In the centre is the remains of the iron style or pin, which served to cast the shadow.[3] (2) On the top of this instrument is a short pillar on the upper surface of which is an amplitude instrument, but it is only divided into octants. Its diameter is two feet two inches.[4] (3) On the level of the terrace is another amplitude instrument, divided into sixty equal parts. Its diameter is only thirteen inches. (4) On the same terrace is a circle, in the plane of the horizon, with a gnomon similar to that of a horizontal

---

† *Mathurā: District Memoir*, 1883, p. 131.

[1] *As. Res.*, V., p. 180.

[2] The fort was built by Mān Singh.

[3] This description is interesting as it corresponds to the instructions given by Jagannāth (*see p.* 39). The Benares dial is similar and possibly had only a north face originally.

[4] There is a similar pillar on the Jaipur instrument (*see plate* XX, *figure* 53), and also one on the top of the gnomon of the Samrāṭ Yantra at Delhi.

dial, but the divisions are equal, and of six degrees each. It must therefore have been intended for some other purpose than the common horizontal dial, unless we may conceive it to have been made by some person who was ignorant of the true principles of that instrument. This could not have been the case with Jaysinha and his astronomers; but the instrument has some appearance of being of a later date than most of the others: they are all of stone or brick, plaistered with lime, in which the lines and figures are cut; and the plaister of the instrument, though on a level with the terrace, and consequently more exposed to accidents than the others, is the freshest and most entire of them all. (5) On the east wall, but facing westward, is a segment, exceeding a semi-circle, with the arch downwards. It is divided into two parts, and each of these into fifteen divisions. Its diameter is four feet. On the west wall, facing eastwards is a similar segment, with arch upwards divided into the same way as the former. Its diameter is seven feet nine inches."

CHAPTER XII.—HISTORICAL PERSPECTIVE.

63. To enable us to place the material collected in its proper historical perspective, it is necessary to survey briefly the development of astronomical science as it affected Jai Singh's work. In making such a survey it is necessary to bear in mind not only the particular theories, topics and methods to be elucidated, but also the views of previous writers.

Of Jai Singh's theories we have but little information: tradition or mistake has allotted to him the whole Ptolemaic theory, and possibly he accepted it all; but he must have been acquainted with the teaching of Kepler, Galilei and Newton for he possessed the works of La Hire, Flamsteed and others. The topics he dealt with are outlined in the preface to the *Zīj Muhammad Shāhi*. Principally, he was concerned with the design of instruments and practical observation with a view to the preparation of a catalogue of the stars, etc. His bent was practical and he was particularly anxious to eliminate instrumental errors.

These points have been illustrated in the foregoing chapters which also have indicated incidentally the sources from which Jai Singh obtained his astronomical notions and inspiration for his methods. There is not the slightest doubt as to the main influence that directed his activities—it was that of the Muslim astronomers of the type of Ulugh Beg; but it is still popularly supposed that Jai Singh's work was principally, if not wholly, of Hindu origin, and previous writers have helped to strengthen the notion. Sir William Jones was one of the first to give this impression: "The Sanskrit work," he says,[1] "from which we might expect the most ample and important information, is entitled *Chetradersa* or a view of geometrical knowledge, and was compiled in a very large volume by order of the illustrious JAYA SINGHA comprising all that remains on that science in the sacred language of India." At considerable trouble and expense this work was published by the Bombay Government and it turned out to be a Sanskrit translation of Naṣīr al-Dīn al-Ṭūsī's edition of Euclid's *Elements*. Hunter was also misleading in a negative way, and more recently Garrett's (otherwise most excellent) book is somewhat remarkably wrong on historical matters. It practically makes Jai Singh the author of the *Almagest*, and the Hindus the inventors of the astrolabe; and generally gives the impression that Jai Singh's work was wholly of Hindu origin. "He revived Hindu astronomy," it tells us, "and gave such an impetus to its study as had not been known in India since the time of Brahmagupta in the seventh century."[2]

It is necessary, therefore, not only to trace Jai Singh's theory and practice back to their proper sources, but to examine in some detail the possible connexions between his work and the traditional Hindu theory and practice.

---

[1] *The works of Sir William Jones, with the Life of the Author.* By Lord Teignmouth, vol. iii, p. 249. Sir W. Jones was usually wrong on astronomical matters. He emphasises his own mistake in this instance by his caution to others: "Provided," he says, "that the utmost critical sagacity were applied in distinguishing such works."

[2] Pp. 19, 20, 21.

For purposes of exposition it is convenient here to speak of the influence of three schools of astronomy (i) Hindu, (ii) Muslim and (iii) European.[1] Jai Singh was to some extent in contact with all three, and it is a matter of considerable interest to determine the quality and quantity of their influence on him. Although he actually lived in the eighteenth century of our era the influences that directed his activities were mediæval : little advance had been made by the Hindu and Muslim schools for centuries and the advances in Europe were too recent to be fully appreciated.

HINDU ASTRONOMY.[2]

64. There is a certain amount of very interesting mythological astronomy recorded in the Vedas but the earliest formal Hindu astronomical works are the *Jyotisha Vedāṅga* and the *Sūrya Prajñapti*, the latter of which exhibits a strange cosmography (with two suns, two moons, etc.[3]) while both have the crude elements of a scientific astronomy. These works are of considerable historical interest : they show little, if any, Greek influence.

Soon after the beginning of the Christian era the traditional astronomical system in India was largely discarded and the system in vogue in the Greek schools was imported and assimilated. In the *Pañcha Siddhāntikā* of Varāha Mihira we possibly have summaries of two western books—the *Pauliśa* and *Romaka Siddhāntas*, but quite apart from this there is abundant evidence to show not only Greek influence but Greek domination. The representative Indian work that exhibits the astronomy of this period is the *Sūrya Siddhānta*. In its original form this work was probably composed about A.D. 400, and the recension now in use about A.D. 1100. Since then very little attempt at advance has been made. The orthodox still accept the *Sūrya Siddhānta* as authoritative and other works are not essentially different.

Such are the facts but there has been an extraordinary amount of misconception current. According to Hindu tradition the *Sūrya Siddhānta* was composed some millions of years ago.[4] Bailly towards the end of the eighteen century considered that Indian astronomy had been founded on accurate observations made thousands of years before the Christian era. Laplace, basing his arguments on figures given by Bailly, decided that some 3,000 years B.C. the Indian astronomers had recorded observations of the planets *correct* to one *second* ; Playfair[5] supported Bailly's views ; Sir William Jones argued that correct observations must have been made as early as 1181 B.C. ; and so on ; but with

---

[1] The question of Chinese influence has not been considered ; but it is interesting to note that in the seventeenth century the French Jesuits helped the Chinese in their astronomy ; and at Peking a few years ago were several large instruments, supposed to be designed by Father Verbiest, copied from those of Tycho Brahe, and also some Muslim instruments of an earlier date. See G. Forbes *History of Astronomy*, pp. 75-77.

[2] The following notes attempt to give very briefly only a fair notion of Hindu astronomy. No attemp has been made at completeness. For further information reference should be made to the works enumerated in the annexed bibliography (p.141 seq.).

[3] The astronomical notions of the early Christian writers were often far more absurd. See p.

[4] *ūrya Siddhānta*, i. 2-3.

[5] Afterwards both Laplace and Playfair recanted. See my *Hindu Astronomy*.

the researches of Bentley, Colebrooke, Weber, Whitney, Thibaut and others more correct views were introduced; and it has long been known that the figures used by Bailly are comparatively modern.

Vedic Astronomy.

65. Vedic Astronomy is more poetical than exact and it is of interest, apart from its poetic value, chiefly as a subject of controversy. Certain scholars, *e.g.*, Dikshit, Tilak, Jacobi and others, argue from rather vague astronomical premises, partly based on the texts, an extreme antiquity for the Vedic writings; others do not accept their views.

The Vedic year was 12 months of 30 days each and an occasional intercalary month, "the thirteenth month fabricated of days and nights, having thirty members." (A.V. XIII, 3, 8.) There is no indication of any definite cycle. (The five-year cycle appears later.) The year was also divided into two equal courses or *ayanas*, a nothern course or *Uttarāyana* beginning at the winter solstice, and a southern course or *Dakshiṇāyana* beginning at the summer solstice.

In the Rig Veda two asterisms only are mentioned, Maghā and Phalgunī; but in later Vedic texts (*e.g.*, A.V., XIX, 7, 1-5) a complete list of the 28 *nakshatras* or asterisms is given. This list is headed by *Kṛittikās* or the Pleiades, which marked, it is believed, the vernal equinox of the Vedic year; and this is a foundation, although not a very secure one, for Vedic chronology.[1]

If the vernal equinox was marked by *Kṛittikās* then the period of fixing this was about 2350 years B.C., when the vernal equinox was approximately of the same longitude as Alcyone (η Tauri) the brightest of the Pleiades. But the only evidence we have is the occurrence of the list of nakshatras with Kṛittikās at the head and, if the assumption made is true, the only legitimate conclusion is that this list must have been prepared at some time *after* 2350 B.C. It may have been an exotic list; or it possibly might be a genuine record of Hindu observation at some time or other. There is another difficulty in the fact that according to the Hindu records Kṛittikās apparently marked the vernal equinox for a very considerable period.

Other parts of the Vedic texts have also been used for the purpose of establishing their great antiquity, *e.g.*, Jacobi attempted to prove that the Vedic year commenced with the summer solstice. His arguments are based on the following very doubtful rendering of a verse of the 'Frog Hymn':—

> "Those leaders of rites observe the institutes of the gods, and disregard not the season of [the twelfth month]:[2] as the year revolves and the rains return, then scorched and heated they obtain Freedom."

Dikshit,[3] from a passage of the Brahmaṇas (Ś.B. II, 1, 2 $^{2-4}$), fixes the age of its composition at 3000 B.C. The words "They (the Kṛittikās) do not move from the eastern quarter while the other asterisms do move from

[1] According to Alberuni (II, 8) in his time the year was commenced with Chitra, Bhādrapadā, Kṛittikā, or Mrigaśiras according to locality or prediliction. See also Fleet *J.R.A.S.*, 1916, p. 570.

[2] Other translations give "the twelve months."

[3] *Indian Antiquary* 1895, XXIV, p. 245.

the eastern quarter" he takes to mean definitely that the asterism Kṛittikas (Pleiades) and no other asterism was on the equator; and he writes "In my opinion the statement conclusively proves that the passage was composed not later than 3000 B.C." Many other similar interpretations have been strived after.

In Vedic texts no definite mention is made of the planets, although much ingenuity has been exercised in interpreting the texts otherwise. There are possible references to eclipses which Ludwig, with some skill, has attempted to identify.

Vedāṅga Astronomy.

66. (*a*) The *Jyotisha Vedāṅga*[1] and the *Sūryaprajñapti* contain the earliest formal astronomical statements. The former introduces the 5-year cycle of 1830 apparent solar days. The year was tropical in theory and contained 366 apparent solar days and was therefore too long. The sidereal year was 367 sidereal days. The lunar day or *tithi* was $\frac{61}{62}$ of an ordinary day, but was reckoned as equivalent to an ordinary day for calendar purposes, one *tithi* being omitted as occasion required. There were 27 nakshatras each supposed to occupy $\frac{360°}{27} = 13\frac{1}{3}$ degrees of the ecliptic, and each nakshatra was considered to be divided into 124 equal divisions or *aṁśas*. The sun therefore traversed 5×27 asterisms in the five year cycle or $\frac{5 \times 27 \times 124}{1830}$ aṁśas a day; and to traverse one nakshatra it took $13\frac{5}{9}$ days $=13$ days 335 *kalās*, since there are 603 *kalās* in one day. The moon traversed $\frac{67 \times 27}{1830} = \frac{603}{610}$ nakshatrās in one day and one nakshatra in 1 day 7 kalās.

The five-year cycle appears to have commenced with the winter solstice and Śravishṭha is said to have marked the beginning of this cycle, and also the beginning of the sun's progress, and also the winter solstice—all of which are in agreement. If Śravishṭha is to be identified with β, α, γ and δ Delphini (as it usually is) then it marked the winter solstice about B. C. 1100. But a list of asterisms given in the text begins with Aśvini (β, and γ Arietis), which marked the vernal equinox about the beginning of the Christian era. The Vedāṅga states that during the northern progress of the sun the days increase in length at an even rate of 1·57 minutes a day or 4 hours 48 minutes in six months of 183 days[2]: the northern and southern progress are considered equal.

(*b*) *The Sūryaprajñapti* is a Jain treatise on astronomy that is similar in many respects to the Vedāṅga. The Jainas held the old Indian idea of the heavenly bodies revolving round mount Meru and as a corollary to this they conceived two suns, two moons and two sets of constellations. The five-year cycle began with the summer solstice with the sun in Pushya and Thibaut thought[3] this was a correction, from actual observation, of the older Vedāṅga.

[1] *Vedāṅga* is the name of certain works or classes of works regarded as auxiliary to the Veda. They are generally considered as of a somewhat later date.

[2] The table of climates on page 131 shows that an increase of $4\frac{1}{2}$ hours corresponds to a latitude of 36° 38′ N., the obliquity being taken as $23\frac{1}{2}$ degrees. For a greater obliquity it would be further north.

[3] Thibaut calculates a precession of 17° 3′ or a difference of 1246 years between the Vedāṅga and the Sūryaprajñapti but gives a caution as to the uncertainty of the deduction.

Another point of difference was the employment of 28 nakshatras of unequal extent and this altered, theoretically, the positions of the nakshatras—in some cases to a very considerable extent, and makes our estimations of the periods in which these works were composed very uncertain.

(c) The characteristics of this period are :—

(1) The five-year cycle.

(2) The division of the sphere into 27 or 28 nakshatras.

(3) Equal daily change in the length of the day.

(4) Omission of any explicit reference to the planets.

Greek Astronomy in India.

67. Varāha Mihira and others, about A. D. 550, made popular new ideas borrowed from the west: they remodelled the Hindu astronomical system on Greek lines. Varāha Mihira's astrological works contain numerous Greek technical terms and show, unmistakably, Greek influence. His great astronomical work, the *Pañchasiddhāntikā* consists of summaries of the Paitāmaha, Vāsishṭha, Romaka, Pauliśa[1] and Saura Siddhāntas. "The Siddhānta made by Pauliśa is accurate, near to it is the Siddhānta proclaimed by Romaka, more accurate is Sāvitra and the two remaining ones are far from the truth." The summary of the *Paitāmaha Siddhānta* exhibits the teaching of the Vedāṅga stage but adds the epoch of 2 Śaka (=A. D. 80). The *Vāsishṭha Siddhānta* appears to represent the transition stage. It alters the longest day rule and introduces shadow calculations and the *lagna* or 'rising sign' notion; while the other three introduce, unequivocally, the Greek teaching. The main characteristics of the *Romaka Siddhanta* are—

(a) A cycle of 2850=19×150 years, perhaps based on the Metonic cycle.

(b) A year of $365^d\ 5^h\ 55^m\ 12^s$ which is exactly the tropical year of Hipparchus.

(c) The epoch of 427 Śaka (=A.D. 505).

(d) Omission of mention of epicycles.

In the *Pauliśa Siddhanta* the following points are noteworthy :—

(a) A year of $365^d\ 6^h\ 12^m$.

(b) A special rule for finding the place of the moon.

(c) Very rough rules for eclipses.

(d) Differences in longitudes between Avanti (Ujjain) and Vārāṇasī (Benares) and Yavana (Alexandria) are given.

(e) A table of sines agreeing with Ptolemy's table of chords.

68. The *Sūrya Siddhānta* is probably the best known astronomical work of the Hindus. The several sections of the accepted text[2] relate to :—

1. The mean motions of the planets.
2. The true places of the planets.
3. The gnomon.
4. Eclipses.[3]

[1] Alberuni writes (i, 153): "Paulisa-siddhānta, so called from Pauliśa the Greek, from the city of Samtra, which I suppose to be Alexandria."

[2] Varāha Mihira's summary of this work differs in some details from the text now in use but not essentially, *e.g.*, the length of the sidereal year in the two works is: Old Sūrya Siddhānta $365^d\ 6^h\ 12^m\ 36^s$. Modern Sūrya Siddhānta $365^d\ 6^h\ 12^m\ 36{\cdot}56^s$. The text now accepted possibly dates from about A.D. 1100 while the earlier limit for the original Sūrya Siddhānta is about A.D. 400.

[3] "That subject which is the greatest mystery, which perplexes the minds of writers of astronomical works." (P.S. i, 5.)

5. Planetary conjunctions.
6. Asterisms.
7. Heliacal risings and settings.[1]
8. Instruments.
9. Time. Cosmogony.
10. Astrology.

69(*a*). The topics dealt with in most of the later Hindu works are fundamentally the same as those of the *Sūrya Siddhānta*, and the following notes apply fairly generally to all these works. The earth is considered as a fixed unsupported sphere round which the other heavenly bodies revolve.[2] Its diameter is given as 1,600 yojanas and the distance of the moon as 51,570 yojanas or roughly the same relative distance as Ptolemy gives (61⅛ radii of the earth). The distances of the other planets are calculated on the assumption that they move with equal velocities. The equation of the centre of a planet is calculated by assuming epicycles, but an apparently indigenous notion is introduced by making the epicycle oval. Ptolemy's theory of the equant is omitted and certain other improvements of Ptolemy relating to the moon and Mercury are also omitted. The precession of the equinoxes is explained as a sort of libration within limits of 27 degrees east and west of a fixed position, at a rate of 54 seconds a year; and the obliquity is generally reckoned at 24 degrees.

The Greek names of the planets and of the signs of the zodiac were adapted and the seven day week introduced; many Greek astrological terms and some Greek mathematical terms were adopted without change. Some of the old cosmological notions that did not interfere with the new ideas were retained. The Greek teaching was, indeed, accepted as a whole — but the evidence points to the curious fact that the Greek astronomy introduced is that of a period preceding Ptolemy, although Ptolemy lived in the second century of our era and the Hindu-Greek astronomical works did not appear earlier than A.D. 400.

69(*b*). One of the most notable features of Hindu astronomy of this period is the employment of immense cycles. To express the planetary elements in integral numbers the astronomers assumed an artificial epoch of general conjunction and a period of recurring conjunctions. The last general conjunction was supposed to be at 3102 B.C. and the cycle or yuga of recurring conjunction is supposed to consist of 4,320,000 years.[3] The planetary elements are expressed in terms of this cycle as follows :—

Planet.	Revolutions.	Planet.	Revolutions.
Sun	4,320,000	Saturn	146,568
Mercury	17,937,060	Moon : Sidereal revolutions .	57,753,336
Venus	7,022,376	„ Synodic „ .	53,433,336
Mars	2,296,832	„ Revolution of apsis .	488,203
Jupiter	364,220	„ „ „ node .	232,238[4]

[1] For some historical account of this topic see Bouché-Leclercq *L'Astrologie grecque*. p. 111*f*.

[2] Āryabhaṭa attempted to revive the theory of Heraclitus that it was the earth's own revolution that produced the apparent motions of the heavenly bodies; but he was condemned as unorthodox and the view has never been accepted by Hindu astronomers generally.

[3] The number 4,320,000 = 20 × 60 [3] is suggestive of Babylonian influence. See also J. Adam *The Nuptial Number of Plato*.

[4] As Whitney points out all these numbers except the last two are divisible by four and this seems to indicate that the last two were later additions.

There was, of course, no general conjunction at 3102 B.C., which period was arrived at by calculating backwards, according to the rules. Bentley assumed that, at that particular time when the calculation was made, the positions of the planets were known with some accuracy: he then calculated, according to the *Sūrya Siddhānta* rules, the positions of the planets at several periods, and found them most correct for A.D. 1091, and concluded that the elements were fixed and the work composed towards the end of the eleventh century.[1]

Hindu Astronomical Calculations.

70. The important elements in the Sūrya Siddhānta are :—

A	Years in the *yuga* or age	4,320,000	A.
B	Sidereal days	1582,237,828	B.
C	Natural or civil days	1577,917,828	C = B—A.
D	Solar months	51,840,000	D = 12 × A.
E	Sidereal months	57,753,336	E.
F	Synodic months	53,433,336	F = E—A.
G	Intercalary months (*adhimāsa*)	1,593,336	G = F—D.
H	Lunar days (*tithi*)	1603,000,080	H = 30 × F.
I	Omitted lunar days (*tithi kshaya*)	25,082,252	I = H—C.

(*a*) One of the most frequent of the Hindu calculations is concerned with finding the *ahargaṇa*, or 'sum of days' that have elapsed since the beginning of a yuga. Thus at the commencement of the Śaka year 953 (A.D. 1031, February 25th) 3,244,132 years of the chaturyuga had elapsed, and to find the number of civil days the calculation is as follows :—

$12 \times 3,244,132 \times \frac{G}{D}$ gives 1,196,525 intercalary months,[2] whence the lunar days elapsed are 30 (12 × 3,244,132 + 1,196,525) = 1,203,783,270.

Again 1,203,783,270 × $\frac{I}{H}$ gives 18,835,679 omitted lunar days, and the number of civil days that have elapsed is therefore

1,203,783,270—18,835,679 = 1,184,947,591.

(*b*) Another characteristic example of the calculations is concerned with the equation of the centre (Kendra). The mean position of a planet was calculated from its number of revolutions in a yuga (p. 74). This was corrected by hypothecating certain epicyclic motions. The mean motion in a circle (deferent) gave wrong positions, so the planet was supposed to move in a second circle (epicycle), whose centre lay on the circumference of the mean circle (deferent), and the corrected position was calculated on this theory. For all but the sun and moon (which required only one) two corrections were made, (1) the equation of the conjunction, (2) the equation of the apsis—by two separate

[1] For further details see my *Hindu Astronomy*. Bentley's general argument is quite sound, but he must not be accepted as reliable on all points.

[2] An intercalary month only occurs when it is complete, hence the fractional part is omitted.

epicycles ; and by combining these two equations the 'true place' of the planet was determined. The calculation is complicated, and entailed very considerable skill of sorts. Here are calculations for the planet Venus for a particular position.

CALCULATIONS FOR THE EQUATION OF THE CENTRE FOR VENUS.

Given mean longitude, L = $8^s\ 18^\circ\ 13'$
Longitude of conjunction, C = $10^s\ 21^\circ\ 50'$
Longitude of apsis, A = $2^s\ 19^\circ\ 52'$
Epicycle of apsis, E, varies from 11° to 12°. Difference, $\Delta = 1^\circ$
Epicycle of conjunction, E, varies from 260° to 262° Difference, $\Delta = 2^\circ$

——	First calculation for equation of conjunction.	Second calculation for equation of apsis.	Third calculation for equation of apsis.	Fourth calculation for equation of conjunction.
Longitude	$8^s\ 18^\circ\ 13'$	$+\frac{26^\circ\ 7'}{2} = 9^s 1^\circ 17'$	$\frac{+22'}{2} = 9^s\ 1^\circ 28'$	$8^s\ 18^\circ 13' + 23' = 8^s\ 18^\circ 36'$
a. Mean commutation = $C-L$.	$2\ 3^\circ 37'$	..	..	$2^s\ 3^\circ 37' - 23' = 2^s\ 3^\circ\ 14'$
a′. Equated anomaly = $A-L$.	..	$5\ 18^\circ 35'$	$\frac{-22'}{2} = 5^s\ 18^\circ\ 24'$	..
b. *Bhujajyā* = Sin a or a' .	3080′	689′	691′	3069′
c. *Koṭijyā* = Cos a or a' .	1527′	3369′	3368′	1521′
d. Correction for epicycle = $\frac{b\Delta}{r}$.	1°47′	12′	12′	..
e. Corrected epicycle = $E-d$.	260°13′	11°48′	11°48′	260°13′
f. $= \frac{be}{360^\circ}$.	2226′	22·3′	22·6′	2218′
g. $= \frac{ce}{360^\circ}$.	1104′	110′	110·4′	1118′
h. *Chala karṇa* = $\sqrt{(r \pm g)^2 + f^2}$.	5058′	3548′	3458′	5067′
i. $\frac{rf}{h}$.	1515	21·0′	22·5′	1505′
j. $= \sin^{-1} i$.	+ 26°7′	+ 0°22	+ 0°23′	+ 25°59′
The true longitude is therefore .		$8^s\ 18^\circ\ 36'$	$+25^\circ\ 59'$	$= 9^s\ 14^\circ\ 35'$

The radius(r) is supposed to be divided into 3438 equal parts, which are, as a matter of convenience, termed minutes, and the sines, etc. (*bhujajyā*, *koṭijyā*, *chala karṇa*, etc.), are expressed in terms of this radius (see my *Indian Mathematics*, p. 11).

The corrections for epicycles are simple proportions : the difference for 90° is Δ, what is the difference for b degrees ? The dimensions of the epicycle are expressed in degrees, etc., in such a way that (r' being the radius of the epicycle) $2\pi r' : 2\pi$ :: the number of degrees : 360.

71. A great many interesting topics must be omitted in this brief sketch of Hindu astronomical theory. Āryabhaṭa taught that the earth rotated upon its axis, and a proper explanation of eclipses, but was not approved. The works of Brahmagupta and Bhāskara have considerable interest in matters of detail, but do not differ fundamentally from the *Sūrya Siddhānta*. Indeed, since the time of composition of this work there has been practically no alteration of fundamental importance in the Hindu theory.

At the present time there are three schools of astronomers : (i) The Saura-paksha, (ii) the Ārya-paksha, (iii) the Brahma-paksha ; and these only differ [1] in matters of detail. For example, a distinctive feature is the length of the year [2] employed. These are :—

(i) *Saura-paksha* year $365^d\ 6^h\ 12^m\ 36{\cdot}56^s$. (ii) *Ārya-paksha* year $365^d\ 6^h\ 12^m\ 30^s$.
(iii) *Brahma-paksha* year $365^d\ 6^h\ 12^m\ 30{\cdot}915^s$.

[1] Really the differentiation is a geographical one. The *Sūrya Siddhānta* is the standard authority in the greater part of India, but the first *Ārya-Siddhānta* is the authority in the Tamil and Malayāḷam countries of Southern India, while Brahmagupta is followed in Gujarāt, Rājputāna and North-West India.

[2] Theoretically, at least, the year is a sidereal one, but there is some vagueness, and there are no records of the methods by which the results were attained. See R. Sewell, *The Indian Calendar*, pp. 7-10.

Hindu Star Lists.

72. The determination of the position of the stars with exactitude does not seem to have interested the ancient Hindu astronomers.[1] In early works the brief lists of stars with celestial co-ordinates given are generally in connexion with the path of the sun and moon through the nakshatras. In each nakshatra the position of a junction star or *yogatārā* was determined. The *Pañcha Siddhāntikā* mentions seven of these while the *Sūrya Siddhānt* gives the position of 28, one for each nakshatra and also of seven other stars. The *Siddhānta Śiromaṇi* and *Brahma Siddhānta* mention only Canopus and Sirius. After these such lists as Mahendra Suri's (given in Appendix A) sometimes occur.

2. The *Pañcha Siddhāntikā* record is as follows :—

> "The yogatārā of **Kṛittkā** is at the end of the sixth degree and three and a half *hastas* to the north of the ecliptic ; that of **Rohiṇī** is at the end of the eighth degree, and five and a half *hastas* to the south of the ecliptic." (XIV, 34) "The two stars of **Punarvasu** are at the eighth degree, and to the north and south of the ecliptic at an interval of eight *hastas*. The star of **Pushya** is at the fourth degree, three and a half *hastas* to the north." (35)
>
> "Of **Āslesha** the southern star is at the first degree one *hasta* (south) ; so also the northern star of **Maghā** the conjunction takes place in its own field,[2] at the sixth degree. Of **Chitra** at seven and a half degrees, three *hastas* to the south."[3]

The *Sūrya Siddhānta* gives the positions of the chief stars of the nakshatras in terms of polar latitude and longitude.[4] The *Sūrya Siddhānta* stars and their position are given in appendix A.

Hindu Astronomical Instruments.

73. The only instruments of practical utility for astronomical purposes described in ancient Hindu works are the sun-dial and clepsydra. An armillary sphere is also described as an instrument for purposes of demonstration. The only Hindu instrument of any antiquity actually found is the clepsydra, consisting of a metal bowl floating in a vessel of water.[5]

The following is a summary of those parts of the early Hindu texts that deal with astronomical instruments. (i) The **Clepsydra** or Water clock is referred to in the *Jyotisha Vedāṅga* where the amount of water that measures a *nāḍikā* (=24 minutes) is given. The more ancient form of water clock appears to have been simply a vessel with a small orifice at the bottom through which the water flowed in a *nāḍikā*[6] but later on there came into use the form described in the *Sūrya Siddhānta* (XIII, 23) : "A copper vessel, with a hole in the bottom

[1] Alberuni writes (II, 83) : "The Hindus are very little informed regarding the fixed stars. I never came across any one of them who knew the single stars of the lunar stations from eyesight, and was able to point them out to me with his fingers."

[2] *i.e.*, on the ecliptic.

[3] Since 24 *aṅgulas* = 1 *hasta* and the diameter of the moon was reckoned as 15 *aṅgulas*, and its mean diameter as 32 minutes (SS. IV, 1) we have approximately 1 *aṅgula* = 2′ 8″ and 1 *hasta* = 51′ 12″ roughly ; but possibly 1 *aṅgula* was meant to measure 2 minutes. Also 27 *nakshatras* occupy 360° and one therefore occupies 13⅓ degrees.

[4] See also p. 8.

[5] It is the only instrument described in the *Āin-i-Akbari* (Ed. Jarret III, 16).

[6] J. F. Fleet *The Ancient Indian Water Clock* J.R.A.S., 1915, pp. 213-230.

set in a vessel of pure water, sinks sixty times in a day and night and is an accurate hemispherical instrument." The *Pañcha siddhāntikā* description[1] (XIV, 32) is similar but adds "Or else a *nāḍikā* may be measured by the time in which sixty slokas each consisting of sixty long syllables can be read out."

A later description of the clepsydra is as follows: "A copper vessel, weighing 10 *palas*, six *añgulas* in height and twice as much in breadth at the mouth —this vessel of the capacity of 60 palas of water and hemispherical in form is called a *ghaṭi*. The aforesaid copper vessel bored with a needle made of 3½ *māshas*[2] of gold and 4 *añgulas* long gets filled in one *nāḍikā*."

In practice, no doubt, the dimensions of the bowl and the orifice were determined by experiment. Bhāskara (XI, 8) indeed says:" See how often it is filled and falls to the bottom of the pail of water in which it is placed. Divide 60 *ghaṭis* of day and night by the quotient and it will give the measure of the clepsydra."

(ii) **The Gnomon.**—The sun-dial described in the early treatises is of the simplest kind, consisting of a vertical rod or gnomon divided into 12 divisions. The descriptions are of a theoretical nature and do not apply so much to the construction of instruments as to theoretical calculations. The *Pañchasiddhāntikā* (XIV, 14-16) instructions are: "Mark from the centre three times the end of the gnomon's shadow and then describe two fish figures. Thereupon describe a circle taking for radius a string that is fastened to the point in which the two strings issuing from the heads of the fish figures intersect and that is so long as to reach the three points marked. On the given day the shadow of the gnomon moves in that circle and the base of the gnomon is the south-north line; and the interval in the north direction is the midday shadow." (III, 1-7) This means mark on any particular day the extremity of the shadow at three different times—and these three points are supposed to lie on a circle the centre of which is found (in the usual way) by the so-called fish figures.[3]

The *Sūrya Siddhānta* directions (III, 1-7) are more elaborate but relate to exactly the same type of dial. They are as follows:—

> "(1) On a stony surface, made water level, or upon hard plaster, made level, there draw an even circle of a radius equal to any required number of digits of the gnomon. (2) At its centre set up the gnomon of twelve digits of the measure fixed upon; and where the extremity of its shadow touches the circle in the former and after parts of the day, (3) there fixing two points upon the circle, and calling them the forenoon and afternoon points, draw midway between them by means of a fish figure, a north and south line. (4) Midway between the north and south directions draw, by a fish figure, an east and west line: and in like manner also by the fish figures between the four cardinal directions, draw the intermediate directions. (5) Draw a circumscribing square by means

[1] Lala Chhotto Lāl's *Jyotisha Vedāñga*, p. 12.

[2] ? About 56 grains troy. Fleet quotes another rule which gives the weight as a *suvarṇa* (= 16 *māshas*) and 4 *añgulas* long drawn out round or square, Bhāskara simply says (XI, 8) it "should have a hole bored in its bottom."

[3] The 'fish figure' is the common part of two intersecting circles.

of the lines going out from the centre: by the digits of its base lines projected upon that is any given shadow reckoned. (6) The east and west line is called the prime vertical (*sama-maṇḍala*): it is likewise denominated the east and west hour circle (*unmaṇḍala*) and the equinoctial circle (*vishuvan maṇḍala*). (7) Draw likewise an east and west line through the equinoctial shadow (*vishuvad-bhā*); the interval between any given shadow and the line of the equinoctial shadow is denominated the measures of the amplitude[1] (*agrā*)."

(iii) **Armillary Sphere.**—The Sūrya Siddhānta (XIII, 1-6) gives instructions for the making of an elaborate armillary sphere :—

(2) "Let the teacher, for the instruction of the pupil ......(3) prepare the wonder working fabric of the terrestrial and stellar sphere (*bhūbha gola*). Having fashioned an earth-globe of wood, of the desired size, (4) fix a staff passing through the midst of it and protruding at either side for Meru; and likewise a couple of sustaining bands and the equinoctial circle; (5) these are to be made with graduated divisions (*aṅgula*) of degrees of the circle (*bhagaṇa*). Further by means of the several day-radii, as adapted to the scale established for those other circles, (6) and by means of the degrees of declination and latitude marked off upon the latter, at their own respective distances in declination, according to the declination of Aries, etc., three (7) bands are to be prepared and fastened: these answer also inversely for Cancer, etc. In the same manner, three for Libra, etc., answering also inversely for Capricorn, etc.: (8) and situated in the southern hemisphere are to be made and fastened to the two band-supporters. Those likewise of the asterisms situated in the southern and northern hemispheres, of Abhijit, (9) of the Seven Sages (*Saptarshayas*) of Agastya, of Brhama, etc., are to be fixed. Just in the midst of all the equinoctial band is fixed. (10) Above the points of intersection of that and the supporting bands are the two solstices (*ayana*) and the two equinoxes (*vishuvat*). From the place of the equinox, with the exact number of degrees, as proportioned to the whole circle, (11) fix by oblique chords, the spaces (*kshetra*) of Aries and the rest; and so likewise another band, running obliquely from solstice to solstice, (12) and called the circle of declination (*krānti*): upon that the sun constantly revolves giving light: the moon and the other planets also by their own nodes, which are situated in the ecliptic (*apa maṇḍala*) (13) being drawn away from it, are beheld at the limit of their removal in latitude (*vikshepa*) from the corresponding point of declination. The orient ecliptic point (*lagna*) is that of the orient horizon; the occident point (*astamgachhat*) is similarly determined. (14) The meridian ecliptic point (*madhyama*) is as calculated by the equivalents in right ascension (*lankodayās*), for mid heaven (*hamadhya*) above. The sine which is between the meridian and the horizon (*kshitija*) is styled the day measure (*antyā*), (15) and the sine of the sun's ascensional distance (*charadala*) is to be recognised as the interval between the equator and the horizon. Having turned upward one's own place, the circle of the horizon is midway of the sphere. (16) As covered with a casing (*vastra*) and as left uncovered, it is the sphere surrounded by Lokāloka. By the application of water is made ascertainment of the revolution of time., (17) One may construct a sphere instrument combined with quicksilver; this is a mystery, if plainly described it would be generally intel-

[1] Distance of the sun at rising or setting from east or west point of the horizon.

ligible to the world. (18) Therefore let the supreme sphere be constructed according to the instruction of the preceptor. In each successive age, this construction, having become lost, (19) is by the sun's favour again revealed to some one or other at his pleasure."

(iv) **Other Instruments**—" So also," the text continues, " one should construct instruments (*yantras*) for the ascertainment of time. (20) When quite alone one should apply quicksilver to the wonder causing instrument.[1] By the gnomon (*sanku*), staff (*yashti*) arc (*dhanus*), circle (*chakra*), instruments for taking the shadow of various kinds .... (21). By water instruments, the hemisphere (*kapāla*), etc., by the peacock, man, monkey, and by stringed sand receptacles, one may determine time accurately (22). Quicksilver-holes, water, and cords, ropes and oil and water, mercury, and sand are used in these: these applications too are difficult .. .. (24). So also a dial (*narayantra*) is good in daytime and when the sun is clear. "

Such is the orthodox Hindu text relating to instruments. Nothing of material value appears to have been added to these instructions until the methods of the *Yavanas* were introduced by Mahendra Sūri and others; but Bhāskara (*Siddhānta Siromani* XI, 16) claims to have invented an instrument called **Phalaka Yantra**[2] which he says is an " excellent instrument, calculated to remove always the darkness of ignorance and is the delight of clever astronomers." This instrument is simply a board divided by horizontals into 90 equal parts. At the centre of the 30th graduation from the top a pin or style is placed perpendicular to the board and round it a circle is drawn of radius=30 divisions, which is graduated in ghaṭis and degrees, and attached to the pin is an index arm (paṭṭika). The instrument is suspended by a chain and is used for observational purposes. It is in fact part of a very simple astrolabe. Bhāskara did not seem very pleased with his instrument for he concludes (XI, 40) by saying " But what does a man of genius want with instruments about which numerous works have treated ? Let him only take a staff in his hand, and look at any object along it, casting his eye from its end to the top. There is nothing of which he will not then tell its altitude, dimensions, etc." This sums up very well indeed the attitude of Hindu astronomers.

Muslim Astronomy.

74. The Muslim astronomers frankly acknowledged their indebtedness to Greek writers. Indeed they were to some extent the direct successors of the Greeks in intellectual matters and the historical problems of their astronomy are much less complicated than is the case with the Hindus. In the middle ages they were the foremost astronomers of the world. They accepted the fundamental features of the Ptolemaic system of the universe. They were aware of the precession of the equinoxes and discovered the slight movement of the apogee of the sun and also they perceived the variation in the obliquity of the ecliptic. They discussed the possibility of the earth revolving on its own axis but generally rejected the theory.

[1] See the *Siddhānta Śiromaṇi* xi 50-51. The instrument appears to be a perpetual motion machine which consists of a wheel with hollow (?tangential) spokes which are filled with mercury. "The wheel thus filled will, when placed on an axis supported by two posts, revolve of itself." This is an old friend.

[2] Compare the 'Balance Khorarie ou Fézaire' described by Delambre *Astronomie du Moyen Age*, p. 5?1.

They fully realised the necessity for methodical observation and in practical astronomy they excelled the Hindus and Europeans of their time.[1] The first series of regular observations with the aid of fairly accurate instruments appears to have been made at Gondeshāpūr in the south-west of Persia in the first years of the ninth century of our era. During the Califiate of al-Ma'mūn (A.D. 813—833) at the observatory at Baghdād all the fundamental elements of the Almagest were verified—the obliquity of the ecliptic, the precession of the equinoxes, the length of the solar year, etc. A measurement of an arc of the meridian in the region of Palmyra was also carried out during the same period, and similar observations continued to be made throughout the Muslim world until the middle of the fifteenth century. The observatory at Cairo was founded in the tenth century, and the observations there were recorded in the 'Hākimid Tables' (al-zīj al-Hākimī). In Persia an observatory was founded in A.D. 1074 at Naisābūr and there in A.D. 1118 al-Khāzinī complied his 'Sanjario Tables' *(al-zīj al-Senjarī)*. In 1259 a great observatory was founded at Marāgha in North-West Persia and there Naṣīr al-Dīn-Ṭūsī (mentioned by Jai Singh) published his famous 'Ilkhānic Tables.'

With Ulugh Beg, the grandson of Tamerlane, the study of scientific astronomy throughout the Islāmic world ceased. He founded a large observatory at Samarqand to which he summoned such renowned astronomers as Jamshīd al-Kāshi (mentioned by Jai Singh), Kādī Zāde, al-Rūmī, 'Alī al-Ḳūsji, and others. He undertook a complete revision of the catalogue of the stars—based upon direct observation—and himself wrote a preface to the tables a few months before he perished by an assassin's hand. Jai Singh professedly followed Ulugh Beg in his astronomical work.

The names of many Muslim astronomers of the middle ages such as Ibn Sīnā or Avicenna, al-Bīrunī, Omar Kahijāmī and Averroēs are familiar to everybody.

75. The practical view taken by the Muslim astronomers led to attempts to improve the instruments in use and to the design of others.

The Marāgha instruments.

A brief description of the instruments used by Naṣr al-Dīn al-Ṭūsī at the observatory at Marāgha is available.[2] The theory of these instruments was probably known to Jai Singh (see p. 4.).

The Marāgha instruments were—

(1) A quadrant or mural circle constructed of wood, the radius of which was about 11 feet. The arc was of copper, 3 inches wide, and was graduated. At the centre was a copper pin round which the alhidade (furnished with two sights) turned. The alhidade terminated in a point and was moved by a cord passing over a pulley attached to the wall.

---

[1] It is related that the works of Āryabhaṭa and Brahmagupta were introduced into Baghdād in the ninth century of our era, and that these works possibly had some influence in directing the scientific study of astronomy by the Arabs; but the 'spirit' of the Arab astronomy is entirely different from that of the Hindu.

[2] For an account of earlier instruments see al-Battāni *Opus astronomicum* LVI-LVII. Ed. C. A. Nallino.

(2) An armillary sphere of five circles, *viz.*, the ecliptic, colure, great circle of latitude, the meridian and the small circle of latitude. The ecliptic, meridian and small circle of latitude were graduated down to minutes and the last mentioned was furnished with an alhidade or sighter. A sighting tube appears to have been used on the alhidade (see figure 68).

(3) A meridian circle of about 11 feet in diameter furnished with an alhidade.

(4) An equatorial circle fixed in the meridian.

(5) An instrument for measuring the diameter of the sun or moon. This consisted of two sights fixed on a bar. The objective was pierced with a comparatively large hole and was moveable along the bar over appropriate graduations. Special discs (like camera stops) were used with the objective.

(6) The instrument of two pillars. At the centre of a bar supported by two pillars a pin was fixed, round which revolved an alhidade or sighter 12 feet long. Vertically beneath this pin was another to which was fastened a graduated bar 17½ feet long and along this bar the end of the sighter was free to run. This appears to have been a modification Ptolemy's parallactic rulers.[1]

(7) A large azimuth circle fixed on a pillar traversed by two diameters directed to the four points of the compass. At the centre were fixed two vertical quadrants furnished with alhidades so that the altitudes and azimuths of two stars could be taken simultaneously.[2]

(8) Sine and Azimuth instrument—An azimuth circle similar to the preceding but in place of the vertical quadrants were two bars moving in a groove and supported by two other bars perpendicular to them.

(9) Similar to No. 6 but with the bars horizontal for measuring azimuths.

The notion of increasing the size of the instrument as far as possible has already been referred to (page 35). It is perhaps to Abu'l Wefa that we owe some of the immense instruments which the Arab works mention. With a quadrant of over 20 feet radius the obliquity of the ecliptic was observed in A.D. 995. The sextant of Abu M. al-Khojendi (C. A.D. 992) was of nearly sixty feet radius. In the tenth century the aperture dial was used and Naṣir-al-Dīn,[3] by utilising a hole in the dome of a large building, obtained excellent results. According to Greaves the quadrant used by Ulugh Beg was 180 feet high.[4]

The astrolabe,[5] the theory of which was due to Ptolemy, was improved by the Muslims almost to perfection and many of these instruments were so

[1] The graduated bar, of course, measures the chords of the arcs, and "as it was much easier to graduate a straight line than an arc, the *triquetrum* continued to be the favourite instrument down to the end of the sixteenth century." (DREYER, p. 332.) This is the *Zāt al Shu' batayn* mentioned by Jai Singh (see page 12).

[2] Nos. 7 and 8 are interesting to us as being the same in principle as the *Rām Yantra* (see page 37).

[3] The *shashṭāmśa yantra* is an aperture dial of this kind.

[4] See L. P. E. A. Sédillot, *Prol. Tab. Astr. d'Ouloug Beg*, pp. lxxx, xcviii, cxxix.

[5] That is the flat astrolabe or *Astrolabium planisphærum*.

beautifully made as to become valuable works of art as well as efficient calculating machines and useful instruments of observation. As a portable instrument of observation it was only superseded about A.D. 1731 by Hadley's quadrant; for purposes of astrological calculations the astrolabe is still in use in the East. The flat astrolabe has already been described in some detail, as also has al-Zarqālī's modification of it and al-Bīrūnī's invention has also been referred to (p. 37). There is also the linear astrolabe or '*asa 'l Ṭūsī* (the rod of al-Ṭūsī),[1] called *yashṭi* by the Hindus. Great ingenuity was exercised in devising improvements and variations of the astrolabe and there are numerous Arabic and Persian works describing the theory and construction of the instrument. The term al-Asṭurlābi as a name suffix was not at all uncommon.[2]

European Astronomy.

76. In Europe, after the death of Ptolemy in the second century of our era, very little advance was made for a thousand years. The Christian church often opposed scientific enlightenment and sometimes persecuted those who sought it; and the patristic writings contain the grossest of astronomical absurdities.

But about the thirteenth century sounder opinions began to prevail and in the early part of the sixteenth century Copernicus wrote his *De Revolutionibus Orbium Cœlestium*. Tycho Brahe, Kepler and Galilei preceded Jai Singh by about a century. Greenwich observatory was founded some forty years before that at Delhi. Newton's *Principia* was written at the time of Jai Singh's birth; Huygens died a few years later; Flamsteed's catalogue of stars was first printed in 1688; Halley in 1705 predicted the return of the comet named after him; the aberration of light was discovered in 1727. Jai Singh succeeded to the Amber territory in 1699 and the Delhi observatory was built about 1724.

77. The European instruments at the beginning of the seventeenth century were in principle much the same as those used by the Greeks and Arabs. Tycho Brahe[3] (1546-1601) had several sextants and quadrants, a parallacticum (see p. 82), and armillary circles; Hevelius (1611-1687) had a somewhat smaller battery of similar instruments; and Flamsteed (1646-1720) used a quadrant of 3 feet, and a sextant of 6 feet radius.

The telescope was used for the general observation of heavenly bodies in 1609 by Galilei and telescopic sights were first systematically used about A.D. 1667. Gascoigne was probably the first (*circa* A.D. 1640) to introduce these and he also invented a micrometer. Some twenty years later Huygens devised, independently, the same contrivance. Hevelius introduced the vernier and tangent screw; Flamsteed used cross wires in the eye pieces of his sighters; Galilei had used a pendulum for short time measurements; Huygens devised a pendulum clock (1656) and Jean Picard (1620-1682) introduced regular time obser-

[1] See *Journal asiatique* 9ᵉ Sér 464*f*, etc.

[2] *E.g.*, 'Alī b. 'Īsā al-Asṭurlābi (9th century), Fath b. Najīyu a'-Asṭurlābi (born A.D. 941—42), Hibetallāh b. al-Ḥosein al-Asṭurlābi (died A.D. 1139-40).

[3] See DREYER's *Tycho Brahe* xii. 215 ff.

vations at the new observatory at Paris.[1] J. D. Cassini (1625-1712), it is stated, devised schemes (never realised) for the construction of gigantic instruments.[2]

78. The Hindus, Arabs and Europeans all derived the fundamentals of their astronomical science from the Greeks. It was the Hindus who first profited by Greek experience, then the Arabs and lastly the Europeans. The last indeed obtained their knowledge of Greek astronomy primarily through the Arabs.

The following chronological scheme exhibits fairly well the relationship between the three schools :—

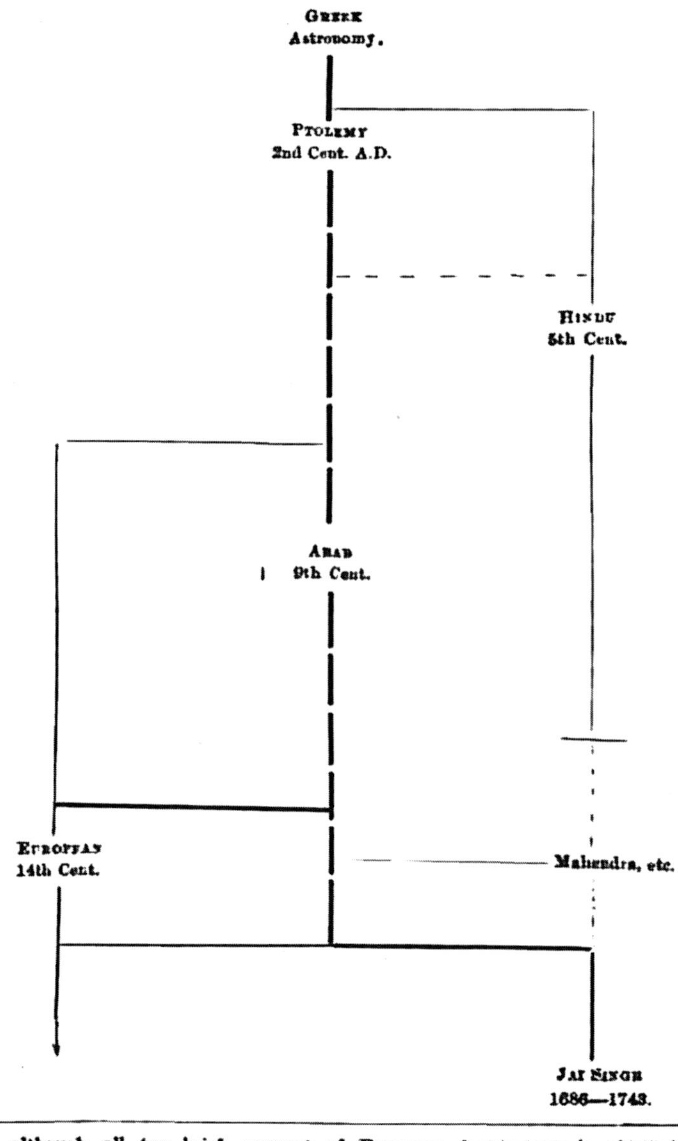

[1] An excellent, although all too brief, account of European instruments is given in the *History of Astronomy* by G. Forbes, while Dreyer's *Tycho Brahe* is the only adequate account of mediæval astronomy (and instruments) in the English language.

[2] A. Berry—*A Short History of Astronomy*, p. 160.

When we examine carefully the methods of the several schools we find somewhat marked differences. Even among the Greek astronomers (*e.g.*, Ptolemy himself) there was a distinct tendency to work on the observations recorded by their predecessors, and in the later Greek schools there was a consequent neglect of observational astronomy. With the Hindus this tendency was emphasised to a remarkable extent and practical work was neglected almost completely. The instruments they describe are either for purposes of what may be termed theoretical calculations or for purposes of demonstration (see page 78). They built no observatories and they made no systematic records of observations.

The Arabs and other Muslim astronomers took an entirely different line. They recognised the value of practical observation; but they hardly suspected the need for a re-examination of the Greek theories. They built observatories and devised improvements in the instruments and set about verifying and correcting Ptolemy's elements.

The European astronomers were perhaps not quite so bound by tradition as were the Hindus and the Arabs. The death of the Ptolematic theory and the invention of the telescope gave a great impetus to research and the European astronomers largely discarded the methods of their predecessors. They recognised the inevitability of observational error and devised means to counteract it; they were forced to consider as of great importance facility of observation and gradually they devised instruments of types unimaginable to their Arabic teachers.

Chapter XIII.—THE EVOLUTION OF JAI SINGH'S INSTRUMENTS.

79. The history of the evolution of Jai Singh's astronomical instruments would, if it could be recorded step by step, be of great interest; but detailed descriptions of the larger Arabic instruments are not generally available and we must for the present be content with general indications of the lines of development. Generally speaking, Jai Singh's instruments are copies of, or direct developments from, those used by Ulugh Beg and his predecessors and successors. The flat astrolabe played an important part. Jai Singh's first attempt at improvement was apparently the construction of huge astrolabes such as those shown in figures 28 and 29 and the construction of large graduated circles. He had some excellent Arabic and Persian models (figures 5—20) but the metal instruments he had constructed do not appear to be of that fine workmanship that adds so much to the value of many of the mediæval astrolabes. As far as can be gathered Jai Singh did not use the ordinary sextant and quadrant instruments such as were used by Naṣīr al-Dīn Ṭūsī, Tycho Brahe, Flamsteed and others.

It has been related how he discarded brass instruments and pinned his faith on large immoveable masonry instruments; and some of these he claims to have devised himself. As already has been pointed out (pages 12, 13) the basic idea was not peculiar to Jai Singh. The Arab, Persian and Tartar astronomers had constructed huge instruments; and they had formulated the notion that the only bar to accuracy of observation was the limit imposed by circumstances on the size of the instruments. Jai Singh was prepared to carry out the idea on which this proposition is based to, what he thought, a reasonable extent.

80. The bases of the designs of the particular instruments are always obvious but Jai Singh showed very considerable ingenuity in the actual constructions. The Jai Prakāś is practically the hemisphere of Berosus somewhat elaborated and the Samrāṭ Yantra may also be considered as evolved from that instrument. This only means, however, that the dial of Berosus was of a very general nature. It consisted of a hemispherical bowl placed with its rim horizontal and in the centre and the same plane as the horizontal edge was fixed a bead whose shadow on the concave surface of the hemisphere marked the trace of the sun's diurnal path. The resemblance to the Jai Prakāś is striking enough but it is doubtful whether Jai Singh had any knowledge of the earlier instrument: he could only have learnt of it from the Muslim astronomers (*e.g.*, al-Battānī who refers to the principle of the instrument). The Jai Prakāś, however, is something more than the bowl of Berosus for it is fully graduated and appears to have been based upon the Muslim instrument known as *al-Maṣātarah*, descriptions of which are found in the works of the Muslim astronomers.[1]

---

[1] See L. A. Sédillot's *Memoire sur les instruments astronomiques des Arabes*, p. 151f.; also a description by al-Barjendi; also Blagrave's *Art of Dyalling*, 1609. One section of Blagrave's book is—"How to make a dyall on a concave hemispheare of a globe two severall waies," and the second way is that of the Jai Prakāś.

The Samrāṭ Yantra might be considered as a section in the plane of the equator of the hemisphere of Berosus and with the bead extended into a line parallel to the axis; but so could any dial be referred to the same origin. In the British Museum are many dials of the seventeenth and early eighteenth centuries constructed exactly on the same principle as the Samrāt.[1] The Samrāṭ instrument is an equal hour instrument or equinoctial as such instruments are often called. The evolution of the equal hour instrument is of considerable interest. In early instruments the time from sunrise to sunset is divided into 12 equal portions. These portions of time vary in length from one-twelfth of the longest day to one-twelfth of the shortest day. They were therefore called unequal hours, also temporal hours and planetary hours (see page 22). Of these varying hours naturally the equinoctial hours were the mean and on the introduction of mechanical clocks the equinoctial hour became the standard for sun-time measurements. According to Delambre, Abdul Hhasan (al-Ḥasan b. ʻAlī b. ʻOmar al-Marrakōshi, Abu ʻAlī) was the first to introduce the constant hour notion among the Muslims; but he seems to have employed, in the usual manner, the horizontal plane for the shadow traces, while Jai Singh's instrument receives the shadow of the inclined gnomon on a circular arc lying in the plane of the equator and thereby secures in the simplest manner equal hours throughout the year. The direct origin of the tangential scales on the gnomon for measuring the declination of the sun is not known; although Ibn Yūnus and other Muslim writers of astronomy had worked out the theory. Hindu astronomers did not employ tangent scales and refer to no other dial than the vertical gnomon and to no other dial measurements than the length of the shadow. They made no direct angular measurements and an angular dial would have been almost contrary to the spirit of their teaching.[3] The only other instruments that can be attributed to Jai Singh's genius are the Digaṁśa Yantra and Rām Yantra but these are simply enlargements in masonry-work of the azimuth and combined azimuth and altitude instruments of the Muslims.

An indicator of the course of evolution of Jai Singh's instruments is still to be seen at Benares on the instrument known as the Chakra Yantra. The wedge (*faras* or 'horse') which fastens the parts of this instrument together is of the traditional Arabic design.

---

[1] *e.g.*, a pocket dial made by Elias Allen about A.D. 1620; a more elaborate dial invented by John Paul Kraus and engraved by T. G. Gutwein; one by Rugendas of Augsburg, 17th century; one by Laurenz Grassl of Augsburg; etc., etc.

[2] Indeed they worked out the complete theory of the horizontal, vertical, inclined, cylindrical and conical dials, etc.

[3] This is a curious point in the history of science. The Hindus seemed to avoid direct angular measurements. Their mathematical works contain no theorems or rules relating to angles (see my *Indian Mathematics*, page 20). Whitney wrote (page 250): "Lest it seem strange that the Hindus should have derived from abroad the name (*koṇa* from γωνία) for so familiar and elementary a quantity as an angle, we would direct attention to the striking fact that in that stage of their mathematical science, at least which is represented by the *Sūrya Siddhānta*, they appear to have made no use whatever in their calculations of the angle."

Chapter XIV.—CONCLUSION.

81. A considerable amount of evidence showing the relationship between Jai Singh's astronomical work and that of his predecessors and contemporaries has been recorded. Let us recapitulate.

The names of the early astronomers and mathematicians referred to in works attributed to Jai Singh are :—

Euclid	*Circa.*	B.C.	290
Hipparchus	"	"	130
Ptolemy	"	A.D.	150
Abdarraḥmān b. 'Omar abū'l-Husain al-Ṣūfī . .	*Died.*	"	986
Naṣīr al-Dīn al-Ṭūsī	*Born.*	"	1201
'Ali b. Muhammad al-Said al-Sharīf		"	1339—1414
Jamshīd b Mes 'ūd Ḡjāt al-Dīn al-Kāshī . .	*Circa.*	"	1440
Ulugh Beg	*Died.*	"	1449
Maulānā Chānd	*Circā.*	"	1550

Of those who came actually into personal contact with Jai Singh the following have been mentioned: Jagganāth, Muhammad Sharīf, Muhammad Mahdi, Padre Manuel Figueredo, Father André Strobel and his companion, Father Claude Boudier and Don Pedro de Sylva.

We know that Jai Singh possessed at least some of works of Ptolemy, Ulugh Beg, P. de la Hire, J. Flamsteed, and also certain European astronomical tables and mathematical text-books. He had Ptolemy's *Almagest* translated into Sanskrit, and a text on the astrolabe compiled, and he brought up to date Ulugh Beg's celebrated catalogue of stars. The instruments themselves are evolved from the types used by the Muslims and Jai Singh's inspiration was avowedly of Muslim origin.

82. The actual points of contact between Jai Singh's astronomical work and that of his predecessors and contemporaries have been generally indicated. Jai Singh himself was a Hindu and had Hindu assistants, the most notable being Jagganāth, who, however, it seems was employed because of his knowledge of Arabic—a somewhat unusual qualification among the Pandits of the day.[1] He refers to one Hindu astronomer by name (see page 11) who was, however, renowned because of his knowledge of Greek methods. Jai Singh was, no doubt, well acquainted with the works of the Hindu astronomers but he does not seem to have made much direct use of them.

Jai Singh had certain Muhammadan assistants (see page 5), he was acquainted with the chief astronomical works of the Muslims, he brought one of their star catalogues up to date, and he copied the instruments of the observatory

[1] There is a tale that Jai Singh was reproached with the statement that the Pandits, who pretended to great learning, were entirely ignorant of Arabic scholarship and he produced Jagganāth, who translated from the Arabic the two great works—Euclid's *Elements* and Ptolemy's *Almagest*. See Ṣudhākara Dvivedi's *Gaṇakataraṅgiṇī* p. 102 f.

at] Samarqand. His masonry instruments were designed after the notions taught by the Muslim astronomers (page 13) and had absolutely nothing in common with those described in Hindu works.

The contact with European astronomical knowledge may not have been really close but it was very definite. Jai Singh sent certain of his assistants to Europe to get books and information; he invited European priests to visit him, and he obtained European tables. There is evidence, however, that his contact with European knowledge was more formal than intimate.

83. We may leave out for the moment the question of European influence, as Jai Singh was really only on the border line of that influence, and consider the Hindu and Arabic schools. The characteristic difference between these is connected with practical work. The Hindus were practical astronomers only in so far as they could calculate from a given starting point with given rules the positions of the planets, eclipses, etc., with some accuracy. This of course implies a very considerable amount of knowledge and skill; but the Hindus had no instruments of precision of their own before Jai Singh's time; neither were they interested in making practical observations of the heavenly bodies. Their rules and the elements given in their approved works sufficed them. The standpoint of the Arabs was entirely different:[1] they were particularly interested in the verification and correction of previously recorded results. They built what were then the finest observatories in the world, and they perfected the astrolabe to an extraordinary degree.

The difference between the two schools is too well known to need elaboration; and the category into which Jai Singh's work places itself is perfectly clearly indicated; and the hypothesis that he received his main astronomical inspiration from Hindu tradition is completely eliminated. He followed "the martyr prince, Mirza Ulugh Beg" of Samarqand. Since both the Hindus and Arabs obtained their astronomy from the Greeks they have much in common but the work of Jai Singh was exactly of that nature which differentiates between the two schools; and what the Muslim astronomers had, and what Hindus lacked, attracted Jai Singh. In his work there is no point of contact with Hindu astronomy that did not also touch the work of the Muslims, while on the other hand there are many points of contact between his work and Muslim astronomy that are remote from the teaching of the Hindu schools.

Jai Singh's apparent indifference to European achievements is rather remarkable; but it must be borne in mind that he very probably only became acquainted with their results after he had conceived and partially carried out his scheme of astronomical research. His tables, it is supposed, were completed about A.D. 1728, and the observatory at Delhi had been built a few years previously. It was in 1728 or 1729 that Jai Singh sent Padre Manuel and others to Europe and in 1734 he was visited by Father Boudier and his companion. These dates might be considered sufficient to account for Jai Singh's neglect of the European discoveries but there is possibly another explanation.

[1] The differentiation between the two schools may be to some extent due to the calendars adopted for religious purposes.

Galilei died a prisoner of the Inquisition in 1642, and his books were not removed from the *Index* until A.D. 1835: Jai Singh's European advisers appear to have been chiefly priests, who, if they were good Catholics[1] would, at that time, have hardly upheld the teaching of Copernicus, Kepler and Galilei! More recent European discoveries might thus have been discredited in Jai Singh's eyes, and he would, at any rate, have found it difficult to reconcile the persecution by authority on the one side with the claim to brilliant scientific discoveries on the other.

84. Jai Singh began his work at a time when European astronomers had arrived at what may be termed the modern conception of the universe. The discoveries of Copernicus, Kepler, Galilei and Newton had been accepted and scientists were settling down to work out in detail the results of their discoveries. Flamsteed's great catalogue was completed just as Jai Singh began his work. But Jai Singh was not in close contact with European ideas and his first astronomical education was probably the study of the work of the Muslim astronomers, particularly Ulugh Beg. In the special circumstances of his experience it is not surprising that Jai Singh refused to follow the lines of research indicated by the European astronomers. Had he done so, his power and his wealth might have enabled him to alter the whole condition of Indian scientific scholarship and, instead of his labours ending with his death when "science expired on his funeral pyre," there might have been established a living school of research. The troubled condition of the country, and the general state of civilization in it, were antagonistic to the progress of science and Jai Singh's work is now only a tradition and his observatories are archæological remains.

That Jai Singh made no new astronomical discoveries is hardly a fair criterion of the value of his work; for, indeed, a great deal of the most valuable astronomical work is not concerned with new discoveries. His avowed object was the rectification of the calendar, the prediction of eclipses and so on—work which entails a great deal of labour and generally shows no remarkable achievement. Considering the state of the country in which Jai Singh lived, the political anarchy of his time, the ignorance of his contemporaries and the difficulties in the way of transmission of knowledge, his scheme of astronomical work was a notable one and his observatories still form noble monuments of a remarkable personality.

---

[1] Condemnation of the correct teaching was not confined to the Roman Catholic Church. See DE MORGAN'S *A Budget of Paradoxes*, in which numerous works opposing the 'Newtonian theory' are quoted.

APPENDICES.

A.—Star Catalogues—

(1) Jai Singh's version of Ulugh Beg's catalogue.

(2) Mahendra's list.

(3) *Sūrya Siddhānta* list.

(4) Stars on the Zarqālī instrument.

B.—Astrological Tables.

C.—Geographical Elements—

(1) Astrolabe Gazetteer.

(2) Some determinations of the positions of Ujjain, Delhi, Jaipur and Benares.

(3) Observatory elements.

(4) Climates and longest days.

D.—Technical terms and symbols, and tables—

(1) Numerical notations.

(2) Signs of the Zodiac.

(3) The planets.

(4) Nakshatras and Manzils.

(5) Obliquity of the ecliptic.

(6) Length of the year.

(7) Precession of the equinoxes.

(8) Hindu measures of time and length.

E.—Chronology.

F.—Bibliography.

APPENDIX A.

Star Catalogues.

A 1. EXTRACTS from the JAIPUR CATALOGUE.

No.	Description as in the MS	Longitude.	Latitude.	Polar longitude.	Declination	Right ascension in Degrees and minutes.	Right ascension in Ghaṭis and palas.	Magnitude.
	CONSTELLATION OF THE LITTLE BEAR 7 STARS.							
		s ° '	° '	s ° '	° '	° '	g. p.	
1.	Star on the edge of the Bear's tail	2 24 23	66 27 N	11 17 0	87 0 N	348 0	58 0	3
2.	Second star on the tail, next to it.	2 26 33	70 0	9 21 10	85 10	293 0	48 50	4
3.	Third star on the tail, next to it.	3 5 3	73 45	8 26 10	82 0	285 0	44 10	4
4.	Star on the left hind foot: 2 stars on the fore-leg, one to the right of the other.	3 21 21	75 36	8 9 0	78 30	247 0	41 10	4
5.	Star on the right hind foot, one to the north.	3 28 23	78 0	8 11 15	76 0	249 30	41 35	5
6.	Star on the left fore-leg, to the South.	4 9 33	73 0	7 17 0	75 0	224 30	37 25	2
7.	Star on the right paw to the north.	4 18 3	75 0	7 25 0	73 45	232 30	38 45	3
1.	Straight to the south from the seventh star.	4 5 3	71 45	7 14 15	76 30 N	221 40	36 57	4
	II. CONSTELLATION OF THE GREAT BEAR.							
1	On the tip of the nose of the Bear.	3 19 3	40 15 N	3 28 30	62 0 N	12(1) 0	20 10	4
2.	First star in the eye of the bear	3 19 57	43 48	4 1 50	65 10	124 30	20 45	5
3.	Second star in the eye of the bear.	3 20 42	43 45	4 3 0	65 5	126 0	21 0	5
4.	Two stars on the forehead: the first of them.	3 20 33	47 54	4 7 0	69 0	130 0	21 40	5
5.	Second star on the forehead .	3 21 51	47 51	4 8 40	68 50	131 40	21 57	5
6.	Star on the ear . . .	3 22 33	51 18	4 14 50	72 0	137 0	22 50	5
7.	Two stars on the neck, the first of them.	3 23 51	44 42	4 8 30	65 0	131 30	21 55	4
8.	Second star on the neck . .	3 26 57	44 54	4 13 0	64 30	136 0	22 40	4
9.	Two stars on the chest, the one to the south.	4 5 27	38 0	4 19 30	55 30	142 45	23 48	4
10.	On the chest, to the north .	4 2 39	42 39	4 20 20	60 30	143 30	23 55	4
11.	Two stars on the knee of the fore-leg: the one to the south.	4 3 30	34 45	4 15 15	52 45	138 0	23 0	3

A 1. EXTRACTS FROM THE JAIPUR CATALOGUE—*concluded.*

No.	Description as in the MS	Longitude.	Latitude.	Polar longitude.	Declination.	Right ascension in		Magnitude.
						Degrees and minutes.	Ghaṭis and palas.	
		s ° ′	° ′	s ° ′	° ′	° ′	g. p.	
12.	Two stars on the paw of the hind leg: the one to the north.	3 29 3	29 21 N	4 26 30	44 5 N	149 30	24 5	3
	* * * * * * * *							
1.	Large brilliant star between the feet.	6 20 39	31 18 N	7 3 0	21 0 N	210 0	35 0	1
	VI. Constellation of the Crown.							
1.	Very brilliant . . .	7 8 38	40 30 N	7 24 0	28 0 N	231 0	38 30	2
2.	Further than this . . .	7 5 48	46 24	7 22 0	30 15	229 5	38 11	4
3.	Above the second to the north .	7 5 18	48 21	7 22 40	32 5	229 50	38 18	4
4.	The third, to the north of this .	7 7 48	50 45	7 26 0	33 15	233 15	38 53	6
5.	Near to the great star to the south.	7 10 26	44 27	7 25 0	27 0	232 15	38 43	4
6.	Near this, a little to the north .	7 12 54	44 42	7 26 30	27 0	233 50	38 58	4
7.	Near to the sixth, to the north .	7 15 3	46 0	7 28 30	28 0	236 0	39 20	4
8.	Near to number 7 . . .	7 14 39	49 30 N	8 0 0	31 0 N	237 30	39 35	4
	VII. Hercules.							
1.	On the forehead . . .	8 12 3	37 9 N	8 16 30	14 30 N	254 30	42 25	3
2.	On the right shoulder . .	7 27 58	42 54	8 6 0	22 0	244 30	40 45	3
3.	On the right arm . . .	7 24 54	39 27	8 3 10	19 0	241 0	40 10	3
4.	On the right side . . .	7 21 57	37 0	8 0 25	17 30	238 0	39 40	4
5.	On the left shoulder . .	8 10 27	47 45	8 16 0	25 0	254 50	42 28	3
6.	On the left arm . . .	8 16 45	49 15	8 20 0	21* 5	259 0	43 10	5
7.	On the left side . . .	8 22 21	51 48	8 25 0	28 0	264 0	44 3	4
8.	In the left palm: three of these to the east.	8 28 54	52 21	8 28 50	28 30	268 30	44 45	4
9.	Of the remaining two, the one to the north.	8 26 33	53 39	8 27 50	30 0	267 30	44 35	4
10.	Of these, the one to the south.	8 25 3	52 39 N	8 26 45	28 50	266 15	44 23	3

*? 27

The above extracts (A1) show the form of the Jaipur catalogue but omit two columns headed respectively *Pārsī Nām* and *Hindu Nām*, as these two columns are mostly blank [see figures 1 and 4]. The essential parts of the catalogue, which are given below (Appendix A1·1), are the longitudes, latitudes and magnitudes, the other columns consisting only of derived elements. The verbal descriptions of the stars are simply translations from Ulugh Beg's Catalogue and the names of the constellations and stars, when given, are mostly transliterations or translations of western names, *e.g.* Sarpa (Draco), Kaikā-us (Cepheus), Silayak (Lyra, Ar. Shilī'ak), Varśava (Perseus), Dalphaina (Delphinus), Trikoṇa (Triangulum), Javvāra (Orion, Ar. Jauzā), Kaitus (Cetus), Naukā (Navis), Muchchhī Yanūvī (Piscis Australis), Aśva mukha (Fam al-Faras), Makara Puchchha (Danab al-Jadi), Iklīla (Corona Borealis, Ar. al-iklīl), Jāt ul-Kurasī (Cassiopeia, Ar. Zāt al-Kursi), Arnava (Lepus), etc., etc. For the unclassed stars (*informatæ*) the expression *Khāriju* (Ar. Khārij) *sūrati* (Ar. Ṣūrat) is used. The term *guchchha* (' a cluster ') is employed to denote a nebula.

In the following table (A 1·1) an asterisk * indicates that there is a discrepancy between Jai Singh's values and those in Baily's version of Ulugh Beg's Catalogue (*Memoirs of the Royal Astronomical Society, 1843*). In the case of the longitudes any difference noted is between Jai Singh's figures and Baily's with 4° 8′ added, this being the amount of precession that had accumulated between the periods of the two catalogues (see page 8).

A table (A 1·2) of differences is added. These are mostly small and do not amount to two per cent. of the whole, and many are obviously copying mistakes.[1] There are indications that the MS. was copied from another Devanāgarī MS., which, in its turn, was copied from one in Persian script. There are numerous examples of what appear to be the result of confusion between the *abjad* symbols (see page 133) for 3 and 8, 4 and 7, tens and thirties, tens and fifties, which confusion is caused by the omission of the dots in the MSS. There are also apparent examples of confusion between the Devanāgarī symbols for one and two. Numbers 360 and 361 appear to have been interchanged ; in 683 Jai Singh's latitude is right, for Baily's value is an emendation. Number 1008 in Baily is omitted (see Baily's note) and the numbers 1009-1019 in Baily correspond to Jai Singh's numbers 1008-1018.

---

[1] The MS. is a good one : that is it is legible and was evidently done with care. It is written on country paper 8·3 × 12 inches in Devanāgarī characters. The copy was made in Samvat 1964.

A. 1·1. The Jaipur Catalogue—Longitudes, Latitudes and Magnitudes.

(This is Ulugh Beg's Catalogue with 4° 8′ added to the longitudes.)

No.	No.	Bayer's Class.	Longitude.	Latitude.	Mag.
Ursa Minor		*(Laghu Balu)*			
			S ° ′	° ′	
1	1	α	2 24 23*	66 27 N.	3
2	2	δ	2 26 33	70 0	4
3	3	ε	3 5 3	73 45	4
4	4	ζ	3 21 21	75 36	4
5	5	η	3 28 23	78 0	5
6	6	β	4 9 33	73 0	2
7	7	γ	4 18 3	75 9 N.	3
Informata					
8	1		4 5 3	71 45 N.	4*
Ursa Major		*(Saptarshayaḥ)*			
9	1	o	3 19 3	40 15 N.	4
10	2	A	3 19 57*	43 48	5
11	3	π	3 20 42	43 45	5
12	4	ρ	3 20 33	47 54	5
13	5	σ^2	3 21 51	47 51	5
14	6	d	3 22 33	51 18	5
15	7	τ	3 23 51	44 42	4
16	8	h	3 26 57	44 54	4
17	9	φ	4 5 27	38 0	4
18	10	υ	4 2 39	42 39	4
19	11	θ	4 3 30	34 45	3
20	12	ι	3 29 3	29 21	3
21	13	κ	3 29 51	29 0	3
22	14	e	3 29 24	36 0	5
23	15	f	3 29 33	33 21	5
24	16	α	4 11 33	49 24	2
25	17	β	4 15 45	45 9	3
26	18	δ	4 27 33	51 30	3
27	19	γ	4 26 39	47 15	3

No.	No.	Bayer.	Longitude.	Latitude.	Mag.
			S ° ′	° ′	
28	20	λ	4 15 18*	29 45 N.	3
29	21	μ	4 (4)7* 15	28 42	3
30	22	ψ	4 24 54	35 15	3
31	23	ν	5 4 15	26 0	3
32	24	ξ	5 4 33	24 45	3
33	25	ε	5 4 39	54 0	2
34	26	ζ	5 12 12	56 12	2
35	27	η	5 23 18	54 9 N.	2
Informatæ					
36	1		5 21 3	40 15 N.	3
37	2		5 14 12	40 39	5
38	3		4 8 9	17 33	4
39	4		4 6 45	19 42	4
40	5		4 9 48	20 18	4
41	6		4 9 18	23 45	4
42	7		4† 3 39	20* 15	6
43	8		3 23 39	23 0 N.	6
Draco		*(Sarpa)*			
44	1	μ	7 21 39	76 15 N.	5
45	2	ν	8 6 48	78 21	4
46	3	β	8 7 9	75 30	3
47	4	ξ	8 23 3	80 0	4
48	5	γ	8 26 3	75 0	3
49	6	b	9 10 58*	82 9	6
50	7	c	9 28 18	78 15	6
51	8	d	9 24 48	80 30*	6
52	9	o	10 14 48	81 24	5
53	10	π	0 11* 9	81 45	3
54	11	δ	0 14 21	83 0	4
55	12	ε	0 29 18	79 9	4
56	13	ρ	0 17 39	77 36	5

No.	No.	Bayer.	Longitude.			Latitude.		Mag.
			S	°	'	°	'	
57	14	σ	0	29	21	80	36* N.	5
58	15	υ	1	17	3	82	0	5
59	16	τ	1	20	42	80	55*	5
60	17	ψ^1	3	8	21	84	12	4
61	18	χ	2	10	3	83	24	4
62	19	φ	2	6	39	84	42	4
63	20	f	4	15	48	87	15	6
64	21	ω	4	4	33	86	45	6
65	22	g	6	2	9	81	57	5
66	23	h^1	6	1	39	84	0	5
67	24	ζ	5	28	42	85	15	3
68	25	η	6	11	3	78	57	3
69	26	θ	6	12	45	74	30	4
70	27	δ	6	1	57	71	27	*4-3
71	28	i	5	1	33	65	21	5
72	29	α	5	4	42	66	27	3
73	30	κ	4	12	45	61	54	3
74	31	λ	4	6	33	57	0 N.	3

Cepheus (*Kaika'us*)

No.	No.	Bayer.	Longitude.			Latitude.		Mag.
75	1	κ	1	29	3	75	45 N.	5
76	2	γ	1	26	39	64	30	4
77	3	β	1	1	45	71	15	4
78	4	α	0	8	42	68	36	3
79	5	η	0	0	33	71	33	4
80	6	θ	0	1	18	73	51	4
81	7	ξ	0	20	18	65	45	5
82	8	ι	0	29	12	62	30	4
83	9	ε	0	10	3	60	0	5
84	10	ζ	10*	11	9	61	15	4
85	11	λ	0	13	3	61	42	6

Informatæ

No.	No.	Bayer.	Longitude.			Latitude.		Mag.
86	1		0	6	18	64	0 N.	5

No.	No.	Bayer.	Longitude.			Latitude.		Mag.
			S	°	'	°	'	
87	2	δ	0	13	33	59	30 N.	4

Bootes (*Avvapurusha*)

No.	No.	Bayer.	Longitude.			Latitude.		Mag.
88	1	κ	5	26	3	58	45 N.	5
89	2	ι	5	27	51	58	51	5
90	3	θ	5	29	12	60	33	5
91	4	λ	6	3	3	54	45	5
92	5	γ	6	14	3	49	24	3
93	6	β	6	20	33	54	27	4
94	7	δ	6	29	25*	49	0	4
95	8	μ	6	29	54	53	27	4
96	9	v^1	6	29	12	57	15	4
97	10		7	1	45	46	27	5
98	11	χ	7	2	39	45	48	5
99	12	c	7	2	3	41	45	5
100	13	ψ	7	0	54	41	21	5
101	14	b	7	1	3	42	48	5
102	15	ω	7	1	36	40	42	5
103	16	ε	6	25	6	40	48	3
104	17	σ	6	20	24	42	9	4
105	18	ρ	6	18	48	42	3	4
106	19	ζ	6	29	27	28	0	4
107	20	η	6	15	51	28	0	3
108	21	τ	6	14	9	26	45	4
109	22	υ	6	15	27	25	0 N.	4

Informata

No.	No.	Bayer.	Longitude.			Latitude.		Mag.
110	1	α	6	20	39	31	18 N.	1

Corona Borealis (*Ikalīla*)

No.	No.	Bayer.	Longitude.			Latitude.		Mag.
111	1	α	7	8	38*	44	30 N.	2
112	2	β	7	5	48	46	24	4
113	3	θ	7	5	18	48	21	4
114	4	π	7	7	48	50	45	6

No.	No.	Bayer.	Longitude.			Latitude.		Mag.
			S	°	′	°	′	
115	5	γ	7	10	26*	44	27 N.	4
116	6	δ	7	12	54	44	42	4
117	7	ε	7	15	3	46	0	4
118	8	ι	7	14*	39	49	30 N.	4

Hercules (*Jāsī*)

No.	No.	Bayer.	Longitude.			Latitude.		Mag.
119	1	α	8	12	3	27*	9 N.	3
120	2	β	7	27	48	42	54	3
121	3	γ	7	24	54	30	27	3
122	4	κ	2*	21	57	37	0	4
123	5	δ	8	10	27	47	45	3
124	6	λ	8	16	45	49	15	5
125	7	u	8	22	21	51	48	4
126	8	o	8	23	54	52	21	4
127	9	ν	8	26	53	53	39	4
128	10	ξ	8	25	3	52	39	3*
129	11	ζ	7	28	18	53	9	3
130	12	ε	8	4	33	53	30	4
131	13	d	8	5	15	55	45	5
132	14	c	8	6	15*	58	36	5
133	15	π	8	8	54	59	51	4
134	16	e	8	10	9	6(0)	15	5
135	17	ρ	8	12	0	60	12	4
136	18	θ	8	24	48	60	51	4
137	19	ι	8	17	3	69	15	4
138	20	x	8	8	21	70	12	6
139	21	y	8	9	57	71	18	6
140	22	z	8	13	18	72	0	6
141	23	η	7	25	3	60	36	4
142	24	σ	7	19	39	63	9	4
143	25	τ	7	10	54	65	48	4
144	26	φ	7	8	45	63	48	4
145	27	υ	7	5	33	64	30	4
146	28	χ	7	5	0	60	15 N.	5

Informata

No.	No.	Bayer.	Longitude.			Latitude.		Mag.
			S	°	′	°	′	
147	1	ω	7	28	21	35	15 N.	4

Lyra (*Aṁgīlī*)

No.	No.	Bayer.	Longitude.			Latitude.		Mag.
148	1	α	9	12	27	62	0 N.	1
149	2	ε	9	15	3	62	30	4
150	3	ζ	9	16*	18	6(0)	15*	4
151	4	δ^2	9	19	3	59	48	4
152	5	η	9	27	18	60	48	4
153	6	θ	9	27	39	59	30	4
154	7	β	9	16	33	56	21	3
155	8	ν^1	9	16	8*	55	15	4
156	9	γ	9	19	15	55	24	3
157	10	λ	9	19	21	54	36	5

Cygnus (*Jāyara*)

No.	No.	Bayer.	Longitude.			Latitude.		Mag.
158	1	β	9	28	33	49	12 N.	3
159	2	φ	1(0)	2	18	50	39	6
160	3	η	10	9	24	54	30	5
161	4	γ	10	22	36	57	51	3
162	5	α	11	2	54	59	42	2
163	6	δ	10	13	15	64	30	3
164	7	θ	10	16	33	69	42*	4
165	8	ι^2	10	16	3	71	6	4
166	9	κ	10	12	48	74	0	4
167	10	ε	10	24	12	49	18	3
168	11	λ	10	26	24	52	0	4
169	12	ζ	10	29	51	43	0	3
170	13	ν	11	2	30	55	0	4
171	14	ξ	11	7	42	56	42	4
172	15	o^1	10	25	36	63	27	4
173	16		10	26	15	64	27*	4
174	17	ω^1	11	6	18	64	21 N.	5

No.	No.	Bayer.	Longitude.			Latitude.		Mag.
Informatæ								
			S	°	′	°	′	
175	1	τ	11	4	51	50	2* N.	4
176	2	σ	11	6	12	51	27	4
Cassiopeia			(*Jālulkurasī*)					
177	1	ζ	1	2	36	43	45 N.	4
178	2	α	1	4	33	46	0	3
179	3	η	1	7	18	46	30	4
180	4	γ	1	10	33	48	30	3
181	5	δ	1	14	27	45	45	3
182	6	ε	1	21	33	46	51	4
183	7		1	24	54	47	36	4
184	8	θ	1	7	45	44	30	4
185	9	φ	1	11	54	44	48	5
186	10	σ	0	26	15	49	30	6
187	11	κ	1	9	33	51	42	4
188	12	β	1	2	9	50	48	3
189	13	ρ	0	27	48	51	0 N.	6
Perseus			(*Varaśavaś*)					
190	1	χ	1	20	27	40	0 N.	Gu
191	2	η	1	25	33	37	9	4
192	3	γ	1	26	39	34	6	3
193	4	θ	1	21	12	31	30	4
194	5	τ	1	24	45	34	0	5
195	6	ι	1	25	48	30	33	4
196	7	α	1	29	15	29	21	2
197	8	σ	1	29	27	27	27	4
198	9	ψ	2	0	51	27	15	4
199	10	δ	2	2	3	26	57	3
200	11	κ	1	24	51	26	0	4
201	12	β	1	23	3	22	0	2
202	13	ω	1	22	48	20	45	4
203	14	ρ	1	21	45	20	21	4
204	15	π	1	20	48	21	9	4

No.	No.	Bayer.	Longtitude.			Latitude.		Mag.
205	16	b	1*	8	54	28	51 N.	4
206	17	λ	2	6	24	28	36	4
207	18	c	2	6	18	25	36	4
208	19	μ	2	7	42	26	19	4
209	20	d	2	8	18	24	45	5
210	21		2	10	15	18	54	5
211	22	ν	2	0	36	21	48	4
212	23	ε	2	2	39	18	54	3
213	24	ξ	2	1	45	14	33	4
214	25	ο	1	28	32	11	30	3
215	26	ζ	2	0	33	10	45 N.	3
Informatæ								
216	1	f	2	5	57	18	54 N.	5
217	2		2	8	51	31	0	5
218	3		1	18	36	20	24	5
Auriga			(*Mamarak ul azinai*)					
219	1	δ	2	26*	30	30	0 N.	4
220	2	ξ	2	26	3	31	0	5
221	3	α	2	28*	51	22	42	1
222	4	β	2	28	0	21	30	2
223	5	ν	2	24	36	14	48	5
224	6	θ	2	26	51	13	33	3
225	7	ε	2	25*	0	20	40	4
226	8	η	2	15	42	18	9	4
227	9	ζ	2	16	3	18	9	4
228	10	ι	2	13	18	10	12	3
229	11	γ	2	19	10	5	15	2
230	12	χ	2	20	48	8	30	6
231	13	φ	2	20	33	10	54 N.	6
Ophiucus			(*Havvā*)					
232	1	α	8	19	12*	35	51 N.	3
233	2	β	8	21	18	28	9 N.	3

No.	No.	Bayer.	Longitude.			Latitude.		Mag.
			S	°	'	°	'	
234	3	γ	8	22	57	25	36 N.	4
235	4	ι	8	6	33	32	33	4
236	5	κ	8	7	48	82*	0	4
237	6	λ	8	2	21	23	48	4
238	7	δ	7	28	33	17	15	3
239	8	ε	7	29	51	16	24	3
240	9	μ	8	20	24	14	45	5
241	10	ν	8	26	27	13	15	4
242	11	τ	8	27	15	14	36	5
243	12	η	8	14	45	6	45	3
244	13	ξ	8	17	12	1	48	4
245	14	A	8	16	48	3	9 N*	4
246	15	θ	8	17	51	2	9N.*	5*
247	16	b	8	18	27	0	18 S.	4
248	17	c^2	8	19	3	0	12 N.*	5
249	18		8	20	27	1	30	5
250	19	ζ	8	6	18	11	45	3
251	20	φ	8	5	12	5	30	5
252	21	χ	8	4	24	3	18	5
253	22	ψ	8	3	54	1	45	5
254	23	ω	8	6	24	0	39 N.	5
255	24	ρ	8	5	15	0	45 S.	5

Informatæ

No.	No.	Bayer.	Longitude.			Latitude.		Mag.
256	1		8	26	48	28	9 N.	4
257	2		8	26	44*	26	15	4
258	3		8	27	12	24	45	4
259	4		8	28	21	26	0	4
260	5		8	29	9	32	21 N.	4

Serpens (*Haiya*)

No.	No.	Bayer.	Longitude.			Latitude.		Mag.
			S	°	'	°	'	
261	1	ι	7	13	9	37	45 N.	4
262	2	ρ	7	15	51	39	42	4
263	3	γ	7	17	42	35	12	3
264	4	β	7	16	21	34	15 N.	3
265	5	κ	7	15	33	37	0	5
266	6	π	7	17	15	42	0	4
267	7	δ	7	15	33	28	45	3
268	8	λ	7	18	36	26	39	4
269	9	α	7	18	33	25	48	3
270	10	ε	7	20	48	24	27	3
271	11	μ	7	22	33	16	15	4
272	12		8	2	48	13	12	5
273	13	ν	8	16	33	10	21	4
274	14	ξ	8	20	48	8	6	4
275	15	ο	8	21	8*	10	36	4
276	16	ζ	8	27	30	19	21	4
277	17	η	9	2	42	20	18	4
278	18	θ	9	12	15	26	56*N.	4

Sagitta (*Sahama*)

No.	No.	Bayer.	Longitude.			Latitude.		Mag.
279	1	γ	10	3	57	39	15 N.	4
280	2	ζ	10	1	42	39	9	6
281	3	δ	10	0	33	38	45	5
282	4	α	9	28	48	39	30	5
283	5	β	9	28	9	38	12 N.	5

Aquila (*Ukāb*)

No.	No.	Bayer.	Longitude.			Latitude.		Mag.
284	1	τ	10	1	39	26	54 N.	6
285	2	β	9	29	33	26	45	3
286	3	α	9	28	18	29	15	2
287	4	ξ	9	29	0	28	33	5
288	5	γ	9	27	21	31	9*	3
289	6	φ	10	0	33	31	0	6
290	7	μ	9	23	12	28	30	6
291	8	σ	9	24	3	26	30	6
292	9	ζ	9	16	39	36	15 N.	3

Informat æ

No.	No.	Bayer.	Longitude. S	°	′	Latitude. °	′	Mag.
293	1	η	9	27	9	21	12 N.	3
294	2	θ	10	1	39	18	27	3
295	3	δ	9	20	24	24	27	3
296	4	ι	9	21	57	19	51	4
297	5	κ	9	23	9	13	39	5
298	6	λ	9	14	27	16	30 N.	3

Delphinus (*Dalphaina*)

No.	No.	Bayer.	Longitude. S	°	′	Latitude. °	′	Mag.
299	1	ε	10	10	30	29	12 N.	4
300	2	ι	10	12	15	28	45	6
301	3	κ	10	11	57	27	36	6
302	4	β	10	12	24	31	45	3
303	5	α	10	13	57	32	51	3
304	6	δ	10	15	3	31	51	3
305	7	γ	10	16	0	32	54*	3
306	8	η	10	11	18	32	12*	6
307	9	ζ	10	11	27	31	21*	6
308	10	θ	10	12	39	30	30 N.	6

Equuleus (*Aśva mukha*)

No.	No.	Bayer.	Longitude. S	°	′	Latitude. °	′	Mag.
309	1	α	10	19	30	20	0 N.	4
310	2	β	10	21	6	20	45	6
311	3	γ	10	19	54	25	0	5
312	4	δ	10	20	48	24	36 N.	5

Pegasus (*Vṛihad aśva khaṁḍa*)

No.	No.	Bayer.	Longitude. S	°	′	Latitude. °	′	Mag.
313	1		0	10	36	25	21 N.	2
314	2	γ	0	5	30	12	24	2
315	3	β	11	25	45	30	51	2
316	4	α	11	20	3	19	0	2
317	5	υ	11	28	3	24	48	4
318	6		11	29	9	24	15	4
319	7	η	11	22	15	34	45 N.	3
320	8	ο	11	21	33	39*	9	5
321	9	λ	11	20	18	28	39	4
322	10	μ	11	21	21	29	0	4
323	11	ζ	11	12	33	17	15	3
324	12	ξ	11	15	21	18	0	4
325	13	ρ	11	16	3	14	15	5
326	14	σ	11	15	6	15	21	5
327	15	θ	11	3	33	15	48	3
328	16	ν	11	2	21	15	15	5
329	17	ε	10	28	36	22	0	3
330	18	π^2	11	15	42	41	0	4
331	19	ι	11	10	27	34	9	4
332	20	κ	11	5	39	36	27 N.	4

Andromeda (*Merāt ulmusalasaloi*)

No.	No.	Bayer.	Longitude. S	°	′	Latitude. °	′	Mag.
333	1	δ	0	18	36	24	0 N.	3
334	2	π	0	19	54	26	5*	4
335	3	ε	0	18	3	22	24	4
336	4	σ	0	17	30	30	45	4
337	5	θ	0	17	24	32	21*	4
338	6	ρ	0	18	48	31	30	5
339	7	ι	0	12	42	41	0	4
340	8	κ	0	13	39	41	49	4
341	9	λ	0	15	0	43	24	4
342	10	ζ	0	17	33	17	18	4
343	11	η	0	19	18	15	36	5
344	12	β	0	27	21	25	36	2
345	13	μ	0	26	6	29	30	4
346	14	ν	0	25	9	32	30	4
347	15	γ	1	10	39	27	36	3
348	16		1	11	3	36	30	4
349	17		1	9	3	35	0	4
350	18		1	5	39*	28	30	4

No.	No.	Bayer.	Longitude. S	°	'	Latitude. °	'	Mag.
351	19	τ	1	5	36	21*	36 N.	4
352	20	ϕ	1	4	3	36	0	5
353	21	A	1	6	48	34	15	5
354	22	χ	1	7	0	31	0	5
355	23		0	4	48	43	42 N.	4

Triangulum (*Musalasitrikoṇamūrttiḥ*)

No.	No.	Bayer.	Longitude. S	°	'	Latitude. °	'	Mag.
356	1	α	1	3	48	36*	6 N.	3
357	2	β	1	9	18	20	15	3
358	3	δ	1	10	15	19	12	5
359	4	γ	1	10	45	18	12	3

Aries (*Mesha*)

No.	No.	Bayer.	Longitude. S	°	'	Latitude. °	'	Mag.
360*	1	γ	1	1	55*	7	51* N.	3
361*	2	β	1	0	21	6	36*	3
362	3	η	1	4	36	7	9	5
363	4	θ	1	5	6	5	16*	5
364	5	ι	1	0	9	5	6	5
365	6	ν	1	11	3	5	45	6
366	7	ϵ	1	14	39	3	12	5
367	8	δ	1	18	3	1	39	4
368	9	ζ	1	19	3	2	30	4
369	10	τ^2	1	20	39	1	39	4
370	11	ρ	1	12	42	1	12	5
371	12	σ	1	11	48	1	24 N.*	5
372	13		1	9	3	5	0 S.	4

Informatæ

No.	No.	Bayer.	Longitude. S	°	'	Latitude. °	'	Mag.
373	1	α	1	4	51	9	30 N.	3
374	2		1	15	9	10	0	4
375	3		1	15	30	12	0	5
376	4		1	13	48	10	54	5
377	5		1	13	3	10	36 N.	5

Taurus (*Vṛisha*)

No.	No.	Bayer.	Longitude. S	°	'	Latitude. °	'	Mag.
378	1	*f*	1	20	28*	6	24 S.	4
379	2	*s*	1	19	57	7	42	4
380	3	ξ	1	18	42	8	54	4
381	4	o	1	18	0	9	39	4
382	5	e	1	24	3	9	0	6
383	6	λ	1	27	51	8	21	3
384	7	μ	2	0	33	12	42	4
385	8	ν	1	27	30	14	45	4
386	9	c^1	2	5	48	9	42	4
387	10	*d*	2	5	21	12	35*	4
388	11	γ	2	3	3	6	9	3
389	12	δ^1	2	3	51	4	9	3
390	13	θ^1	2	5	12	6	15	3
391	14	α	2	6	39	5	15	1
392	15	ϵ	2	5	18	2	54	3
393	16	*i*	2	10	33	4	27	5
394	17	*m*	2	13	24	4	30	5
395	18	*l*	2	13	33	3	0	5
396	19	ζ	2	21	9	2	42 S.	3
397	20	τ	2	8	42	0	30 N.	4
398	21	υ^1	2	4	57	1	0	4
399	22	κ^1	2	4	33	0	9	4
400	23	ω^1	2	0	12	0	39 N.*	6
401	24	ω^2	2	3	12	1	0 S.	5
402	25	*p*	2	1	51	4	48 N.	5
403	26	ψ	2	1	27	6	18	5
404	27	χ	2	4	51	3	33	5
405	28	ϕ	1*	4	33	5	36	5
406	29		1	26	9	3	45	5
407	30		1	26	24	3	30	5
408	31		1	26	57	3	45	5
409	32	η	1	27	6	4	9 N.	4

Informatæ

No.	No.	Bayer.	Longitude.	Latitude.	Mag.
			S ° ′	° ′	
410	1		1 18 51	19 30 S.	4
411	2	ι	2 18* 13*	1 35*	5
412	3	η	2 16 51	1 9	5
413	4	ο	2 18 54	1 30	5
414	5		2 21 42	6 34*	5
415	6		2 22 57	8 0 S.	5
416	7		2 20 24	1 15 N.	5
417	8		2 21 51	2 30	5
418	9		2 23 45	1 48	5
419	10		2 24 21	3 42	5
420	11		2 25 36	2 20 N.	5

Gemini (*Mithuna*)

No.	No.	Bayer.	Longitude.	Latitude.	Mag.
421	1	α	3 16 51	9 54 N.	2
422	2	β	3 20 3	6 30	2
423	3	θ	3 7 33	10 45	4
424	4	τ	3 12 3	7 30	4
425	5	ι	3 15 36	5 30	4
426	6	υ	3 17 57	4 54	4
427	7	κ	3 20 9	2 45	4
428	8	A	3 15 9	2 45	5
429	9	b^2	3 16 3	3* 45	5
430	10	ε	3 6 21	1 15* N.	3
431	11	δ	3 14 51	0 21 S.	3
432	12	ζ	3 11 6	2 18	4
433	13	λ	3 15 6	6 0	3
434	14	η	3 0 3	1 30	4
435	15	μ	3 1 39	1 15	4
436	16	ν	3 3 33	3 24	4*
437	17	γ	3 5 39	7 12	3
438	18	ξ	3 7 39	10 12 S.	4

Informatæ

No.	No.	Bayer.	Longitude.	Latitude.	Mag.
			S ° ′	° ′	
439	1		3* 27 21	0 45 S.	4
440	2	κ	3 0 3	6 0 N.	4
441	3	d	3 8 12	2 0 S.	5
442	4		3 23 9	1 20	5
443	5	g	3 21 21	3 0	5
444	6	f	3 19 54	4 15	5
445	7		3 25 18	2 45 S.	4

Cancer (*Karka*)

No.	No.	Bayer.	Longitude.	Latitude.	Mag.
446	1	ε	4 3 54	1 0 N.	Guchchha
447	2	η	4 1 33	1 21 N.	4
448	3	θ	4 1 48	1 15 S.	4
449	4	γ	4 3 42*	3 6 N.	4
450	5	δ	4 4 51	0 15 S.	4
451	6	α	4 9 48	5 21 S.	4
452	7	ι	4 2 15	10 15 N.	4
453	8	μ^2	3 27 45	0 54 N.	5
454	9	β	4 0 51	10 30 S.	4

Informatæ

No.	No.	Bayer.	Longitude.	Latitude.	Mag.
455	1	o^1	4 8 18	2 15 S.	4
456	2	κ	4 12 3	5 48 S.	4
457	3	ν	4 9 3	5 0 N.	5
458	4	ξ	4 6 57	7 0 N.	5

Leo (*Siṁha*)

No.	No.	Bayer.	Longitude.	Latitude.	Mag.
459	1	κ	4 12 18	10 9 N.	4
460	2	λ	4 14 18	8 0	4
461	3	μ	4 17 33	12 21	3
462	4	ε	4 26 6*	9 45	3
463	5	ζ	4 24 33	11 33	3
464	6	γ	4 26 6	9 45*	2
465	7	η	2* 24 27	4 48	3
466	8	α	4 26 21	0 9 N.	1

No.	No.	Bayer.	Longitude.	Latitude.	Mag.
			S ° '	° '	
467	9	A	4 26 30	1 27 S.	4
468	10	ν	4 24 3	0 12	5
469	11	ψ	4 21 3	0 6	6
470	12	ξ	4 18 30	3 9	6
471	13	ο	4 21* 30	3 57	4
472	14	π	4 25 48	4 0	4
473	15	ρ	5 2 45	0 9 S.	4
474	16	i	5 0 33	4 15 N.	6
475	17	k	5 4 9	5 36	6
476	18	l	5 6 27	2 6	6
477	19	b	5 5 27	13 6	5
478	20	δ	5 7 36	14 9	2
479	21		5 6 48	15* 45	5
480	22	θ	5 9 48	9 24	3
481	23	ι	5 14 6	6 9	3
482	24	σ	5 15 24	1 15 N.	4
483	25	p^5	5 15 39	5 0 S.	4
484	26	ν	5 21 12	3 15 S.	5
485	27	β	5 17 57	12 0 N.	1

Informatæ

No.	No.	Bayer.	Longitude.	Latitude.	Mag.
486	1		4 29 48	14 0 N.	5
487	2		5 2 3	16 30	5
488	3	χ	5 10 51	1 15 N.	4
489	4	c	5 10 27	0 30 S.	5
490	5	d	5 11 24	3 0 S.	5
491	6		5 20 12	28 12 N.	5
492	7		5 20 33	23 30	5
493	8		5 24 36	24 0 N.	5

Virgo (*Kanyā*)

No.	No.	Bayer.	Longitude.	Latitude.	Mag.
494	1	ν	5 20 39	4 39 N.	5
495	2	ξ	5 20 33	6 15	5
496	3	ο	5 24 39	8 24	5*
497	4	π	5 29* 27	6 9	5
			S ° '	° '	
498	5	β	5 22 39	0 10 N.	3
499	6	η	6 1 15	1 30	3
500	7	γ	6 6 21	2 54	3
501	8		6 10 30	3 0	6
502	9	θ	6 14 33	1 36	4
503	10	δ	6 8 9	8 45	3
504	11	ρ	6 1 54	13 30	5
505	12	d^2	6 4 9	11 18	6
506	13	ε	6 5 57	16 15 N.	3
507	14	α	6 20 18	2 9 S.	1
508	15	ζ	6 19 3	8 45 N.	3
509	16	l^2	6 20 24	3 12 N.	5
510	17	h	6 21 27	0 24 S.	6
511	18	m	6 23 3	1 9 N.	5
512	19	i	6 22 33	2 54 S.	5
513	20		6 25 21	1 30 S.	5
514	21	p	6 23 6	8 45 N.	5
515	22	ι	6 29 57	7 15	4
516	23	κ	7 1 0	3 0	4
517	24	φ	7 1 48	11 45	4
518	25	l	7 3 15	0 42	4
519	26	μ	7 6 45	9 51 N.	4

Informatæ

No.	No.	Bayer.	Longitude.	Latitude.	Mag.
520	1	χ	6 8 18	3 42 S.	5
521	2	ψ	6 12 18	3 24	5
522	3	g	6 15 27	3 21	5
523	4		6 20 15	8 0	6
524	5		6 21 27	8 36	5
525	6		6 28 18	7 42 S.	6

Libra (*Tulā*)

No.	No.	Bayer.	Longitude.	Latitude.	Mag.
526	1	α	7 12 15*	0 45 N.	3
527	2	μ	7 10 39	1 45	5

No.	No.	Bayer.	Longitude.			Latitude.		Mag
			S	°	′	°	′	
528	3	β	7	16	6	8	45 N.	3
529	4	δ	7	12	6	8	36	5
530	5	ι^1	7	17	20*	1	46*	4
531	6	ν^1	7	14	54	1	9	5
532	7	γ	7	21	57	4	45	4
533	8	θ	7	26	12	2	57 N.	4

Informatæ

No.	No.	Bayer.	Longitude.			Latitude.		Mag
534	1		7	20	3	8	42 N.	5
535	2		7	26	54	6	30	4
536	3		7	27	33	8	54	4
537	4	λ	7	26	33	0	36	6
538	5	η	7	23	48	3	12 N.	6
539	6	κ	7	24	33	1	24 S.	4
540	7		7	17	27	7	30	3
541	8		7	25	3	8	15	4
542	9		7	25	51	10	0 S.	4

Scorpio (*Vṛiśchika*)

No.	No.	Bayer.	Longitude.			Latitude.		Mag
543	1	β	7	29	30	1	20 N.	3
544	2	δ	7	29	6	2	3 S.	3
545	3	π	7	28	45*	5	27	3
546	4	ρ	7	29	3	8	51 S.	3
547	5	ν	8	0	36	1	45 N.	4
548	6	ω^1	7	29	18	0	30 N.	4
549	7	σ	8	4	36	3	45 S.	3
550	8	α	8	6	24	4	30	2
551	9	τ	8	7	48	6	21	3
552	10	c^2	8	2	21	6	57	5
553	11		8	3	33	7	12	5
554	12	e	8	10	27	12	0	3
555	13	μ	8	12	3	15	55*	3
556	14	ζ^1	8	13	27	18	51	4
557	15	ζ^2	8	13	33	19	15	4

No.	No.	Bayer.	Longitude.			Latitude.		Mag
			S	°	′	°	′	
558	16	η	8	17	9	21*	0 S.	3
559	17	θ	8	22	3	19	21	3
560	18	ι	8	23	36	16	18	3
561	19	x	8	22	39	16	0	3
562	20	λ	8	20	19*	13	33	3
563	21	υ	8	20	3	13	54 S.	3

Informatæ

No.	No.	Bayer.	Longitude.			Latitude.		Mag
564	1		8	24	0*	13	39 S.	4
565	2		8	19	15	6	45	Gu. 5
566	3		8	23	39	4	15 S.	5

Sagittarius (*Dhanu*)

No.	No.	Bayer.	Longitude.			Latitude.		Mag
567	1	γ	8	27	57	7	12 S.	3
568	2	δ	9	1	36*	6	45	3
569	3	ε	8	1	21	11	12	3
570	4	λ	9	2	33	2	0 S.	3
571	5	μ	9	0	0	2	28 N.*	4
572	6	σ	9	8	39	3	45 S.	3
573	7	φ	9	6	27	3	54	4
574	8	ν^2	9	9	15	0	45 S.*	Gu.
575	9	ξ^1	9	9	51	2	0 N.	4
576	10	ο	9	11	39	1	15	4
577	11	π	9	13	3	2	0	4
578	12	d	9	14	57	3	15	5
579	13	ρ^1	9	15	33	4	6	4
580	14	υ	9	16	3	6	15	4
581	15	e^1	9	19	21	5	24	6
582	16	g	9	23	18	6	0	5
583	17	f	9	21	15	1	48 N.	6
584	18	χ^3	9	16	24	1	54 S.	5
585	19	h^2	9	18	48	3	6 S.	4
586	20	ψ	9	13	9	2	18 S.	5
587	21	τ	9	10	39	5	0	4

No.	No.	Bayer.	Longitude.			Latitude.		Mag.
			S	°	′	°	′	
588	22	ζ	9	9	39	7	0 S.	3
589	23	β	9	11	54	22	18	4
590	24	α	9	12	21*	18	36	4
591	25	η	9	0	3	13	18	3
592	26	θ	9	21	3	13	21	4
593	27	ι	9	28*	33	20	39	4
594	28	ω	9	22	15	5	30	5
595	29	A	9	23	3	5	30	5
596	30	b	9	22	33	6	9	5
597	31	c	9	23	15	7	0 S.	5

Capricornus (*Makara*)

No.	No.	Bayer.	Longitude.			Latitude.		Mag.
598	1	α^1	10	0	39	6	42 N.	3
599	2	ν	10	0	57	6	27	5
600	3	β	10	0	18	4	45	3
601	4	ξ^2	9	29	3	7	30	6
602	5	ο	10	1	39	0	42	6
603	6	π	10	1	57*	1	39	6
604	7	ρ	10	1	39	1	21	6
605	8	σ	9	29	21	0	36	6
606	9	τ^2	10	4	30	3	27	6
607	10	υ	10	4	18	0	54 N.	6
608	11	ψ	10	3	33	7	0 S.	4
609	12	ω	10	4	9	8	45	4
610	13	A	10	8	3	8	6	4
611	14	ζ	10	13	24	7	0	4
612	15		10	13	42	5*	12	5
613	16	φ	10	11	3	4	36	6
614	17	χ	10	9	9	4	18	6
615	18	η	10	9	3	2	42	5
616	19	θ	10	10	9	0	0	4
617	20	ι	10	14	3	0	48	4
618	21	ε	10	16	42	5	15	4
619	22	κ	10	18	15	5	0	3*

No.	No.	Bayer.	Longitude.			Latitude.		Mag.
			S	°	′	°	′	
620	23	γ	10	18	21	2	30	*
621	24	δ	10	19	36	2	15 S.	3
622	25		10	19	51	0	15 N.	5
623	26	μ	10	22	18	0	0	5
624	27	λ	10	20	30	2	48	5
625	28	c^1	10	21	42	4	0 N.	5

Aquarius (*Kumbha*)

No.	No.	Bayer.	Longitude.			Latitude.		Mag.
626	1	d	10	24	21	15	15 N.	6
627	2	α	10	29	39	10	9	3
628	3	ο	10	28	42	8	42	5
629	4	β	10	19	51	8	48	3
630	5	ξ	10	20	48	6	45	5
631	6		10	10	15	7	6	6
632	7	μ	10	9	30	8	9	5
633	8	ε	10	7	57	8	9	4
634	9	γ	11	3	27*	8	0	3
635	10	π	11	5	3	10	9	4
636	11	ζ	11	5	15	8	48	3
637	12	η	11	7	30*	8	0	3
638	13	θ	10	29	51	1	48	4
639	14	ρ	11	0	18	2	18 N.	5
640	15	σ	11	2	9	1	15 S.	4
641	16	ι	10	25	45	1	54 S.	4
642	17		10	27	18	4	45 N.	6
643	18	δ	11	6	3	8	18 S.	3
644	19	τ^2	11	5	45	5	45	4
645	20		10	28	51	6	9	6
646	21	g^2	11	21*	42	11	0	5
647	22	g^1	11	1	57	10	6	5
648	23		11	8	39	0	18*	4
649	24	λ	11	6	52*	1	10	4
650	25	φ	11	13	6	0	30	4
651	26	χ	11	13	33	2	0 S.	4

No.	No.	Bayer.	Longitude. S	°	′	Latitude. °	′	Mag.
652	27	ψ^1	11	13	3	3	24 S.	4
653	28	ψ^2	11	13	42	4	0	4
654	29	ψ^3	11	13	27	5	0	4
655	30		11	11	42	8	48	5
656	31	ω^1	11	16	3	11	30	5
657	32	ω^2	11	16	15	11	0	5
658	33	A^1	11	15	3	14	30	5
659	34	i^1	11	15	42	15	6	5
660	35	i^2	11	16	33	15	42	5
661	36	b^1	11	10	51	15	0	4
662	37	b^2	11	11	24	15	54	4
663	38	b^3	11	12	15	16	45	4
664	39	c^1	11	5	21	16	57	4
665	40	c^3	11	16*	12	15	51	4
666	41	c^2	11	6*	23*	14	48	4
667	42		11	0	27	21	24 S.	1

Informatæ

No.	No.	Bayer.	S	°	′	°	′	Mag.
668	1		11	20	48	16	33 S.	4
669	2		11	23	20*	15	45	4
670	3		11	22	36	19	18 S.	4

Pisces (*Mina*)

No.	No.	Bayer.	S	°	′	°	′	Mag.
671	1	β	11	14	54	8	54 N.	4
672	2	γ	11	17	57	7	12 N.	4
673	3	b	11	19	33	8	42 N.	4
674	4	θ	11	21	57	8	48	4
675	5	ι	11	23	57	7	0	4
676	6	κ	11	29*	24	4	0	4
677	7	λ	11	23	30	3	0	4
678	8	ω	11	29	15	6	18	4
679	9	d	0	4	58	5	24	6
680	10		0	6	57	3	0	6
681	11	δ	0	11	3	1	54	4
682	12	ϵ	0	14	39	1	12 N.	4
683	13	ζ	0	17	3	6*	0 S.	4
684	14	e	0	16	30	1	39	6
685	15	f	0	17	3	4	54	5
686	16	μ	0	20	3	2	30	4
687	17	ν	0	22	33	5	0	4
688	18	ξ	0	23	57	8	45	4
689	19	α	0	26	3	9	30	3
690	20	o	0	24	33	2	12 S.	4
691	21	π	0	24	12	1	48 N.	5
692	22	η	0	23	54	5	0	3
693	23	ρ	0	24	18	8	36	5
694	24	g	0	25	30	22	9	5
695	25	τ	0	24	54	21	21	5
696	26	h	0	21	42	20	45	6
697	27	k	0	20	51	19	42	6
698	28	i	0	19	48	20	30	6
699	29	ψ^1	0	20	27	12	51	4
700	30	ψ^2	0	20	36	11	54	4
701	31	ψ^3	0	20	54	10	57	4
702	32	υ	0	25	3	18	0	4
703	33	ϕ	0	23	36	14	45	4
704	34	χ	0	21	18	12	0 N.	4

Informatæ

No.	No.	Bayer.	S	°	′	°	′	Mag.
705	1		11	24	54	3	12 S.	4
706	2		11	25	18	3	0	4
707	3		11	25	33	6	12	4
708	4		11	26	21	6	12 S.	4

Cetus (*Kaitus*)

No.	No.	Bayer.	S	°	′	°	′	Mag.
709	1	λ	1	11	39	8	18 S.	4
710	2	α	1	11	3	12	11*	3
711	3	γ	1	6	18	12	18	3

No.	No.	Bayer.	Longitude. S	°	'	Latitude. °	'	Mag.
712	4	δ	1	4	30	14	42 S.	3
713	5	ν	1	3	54	8	9	4
714	6	ξ^2	1	7	15	6	30	4
715	7	ξ^1	1	2	3	4	24	4
716	8	ρ	0	26	45	25	42	4
717	9	σ	0	27	12	29	15	4
718	10	ϵ	1	0	33	26	15	4
719	11	π	1	0	51	28	51	4
720	12	τ	0	15	3	25	30	3
721	13	υ	0	16	15	31	0	4
722	14	ζ	0	18	45	21	9	3
723	15	θ	1*	13	3	16	15	3
724	16	η	0	8	48	16	42	3
725	17	ϕ^3	0	4	27	15	6	6
726	18		0	2	48	17	12	6
727	19		0	2	48	15	21	5
728	20		0	2	21	16	6	5
729	21	ι	11	28	3	10	30	3
730	22	β	11	29	33	21	0 S.	3

Orion (*Javvara*)

No.	No.	Bayer.	Longitude. S	°	'	Latitude. °	'	Mag.
731	1	λ	2	20	30	13	30 S.	Gu.
732	2	α	2	25	21	16	45	1
733	3	γ	2	17	42	17	45*	2
734	4	A	2	18	48	17	39	4
735	5	μ	2	26	48	14	0	4
736	6	k^2	3	0	24	11	15	6
737	7	ξ	2	29	45	9	15	5
738	8	υ	2	29	12	8	42	5
739	9	f^2	3	0	12	7	15	6
740	10	f^1	2	29	18	7	15	6
741	11	χ^1	2	25	15	3	24	5
742	12	χ^2	2	27	24	3	45	5
743	13	ω	2	11*	3	19	24	4
744	14	n^2	2	20	24	19	42 S.	6
745	15	n^1	2	18	21	20	9	6
746	16	ψ^2	2	17	39	20	30	5
747	17		2	13	48	7	45	4
748	18		2	12	54	7	54	4
749	19	g	2	12	21	10	6	4
750	20	π^4	2	9	48	12	42	4
751	21	π^2	2	8	51	14	18	4
752	22	π^1	2	8	24*	15	30	3
753	23	π^3	2	8	47*	16	45	3
754	24	π^5	2	8	54	20	18	3
755	25	π^6	2	9	57	21	12	4
756	26	δ	2	18	42	23	57	2
757	27	ϵ	2	20	18	24	36	2
758	28	ζ	2	21	12	25	24	2
759	29	η	2	16	3	25	39	3
760	30	c	2	19	21	27	54	4
761	31	θ^2	2	19	27	28	27	3
762	32	ι	2	19	42	29	12	3
763	33	d	2	20	33	30	42	4
764	34	ν	2	29*	39	30	51	4
765	35	β	2	13	33	31	18	1
766	36	τ	2	14	45	30	24	4
767	37	e	2	16	0	31	15	4
768	38	κ	2	22	48	33	21 S.	3

Eridanus (*Kulpā*)

No.	No.	Bayer.	Longitude. S	°	'	Latitude. °	'	Mag.
769	1	λ	2	12	3	31	54 S.	4
770	2	β	2	12	15	28	12	4
771	3	ψ	2	9	48	28*	12*	4
772	4	ω	2	7	21*	27	28*	4
773	5	μ	2	6	9	25	48	4
774	6	ν	2	3	24	24*	24	4
775	7	ξ	1	29	27	26	0	[illegible]

No.	No.	Bayer.	Longitude. S	°	′	Latitude. °	′	M.
776	8	d	1	27	3	28	15	4
777	9	o	1	25	48	27	39	4
778	10	γ	1	20	48	33	15	3
779	11	π	1	17	39	31	15	4
780	12	δ	1	17	18	29	0	3
781	13	ϵ	1	14	54	27	48	3
782	14	ζ	1	10	42	26	9	4
783	15	ρ^3	1	8	15	23	54	5
784	16	η	1	5	24	24	30	4
785	17	σ	1	4	22	24	12	5
786	18	τ^1	0	28	48	33	0	4
787	19	τ^2	0	29	33	35	39	4
788	20	τ^3	1	1	48	30*	45	4
789	21	τ^4	1	7	15	38	30	4
790	22	τ^5	1	10	57	39	27	4
791	23	τ^6	1	14	33	41	30	4
792	24	τ^7	1	14	45	12*	30	5
793	25	τ^8	1	15	9	44	0	4
794	26	τ^9	1	17	18	44	6	4
795	27	υ^6	1	25	51	50	42	4
796	28	υ^7	1	26	18	51	45	4
797	29	υ^5	1	20	30*	54	30	4
798	30	υ^4	1	18	9	54	9	4
799	31	υ^3	1	8	9	54	3	4
800	32	υ^2	1	6	48	55	39	4
801	33	υ^1	1	4	33	55	0	4
802	34	θ	0	19	48	53	45 S.	1

Lepus (*Arnava*)

No.	No.	Bayer.	Longitude. S	°	′	Latitude. °	′	Mag.
803	1	ι	1*	11	48	35	0 S.	5
804	2	κ	2	11	39	36	0	5
805	3	ν	2	14	3	35	30	5
806	4	λ	2	13	51	36	18 S.	5
807	5	μ	2	11	18	39	30	4

No.	No.	Bayer.	Longitude. S	°	′	Latitude. °	′	Mag.
808	6	ϵ	2	8	18	45	30 S.	4
809	7	α	2	17	9	41	18	3
810	8	β	2	15	48*	44	12	3
811	9	δ	2	23	18	44	9	4
812	10	γ	2	20	51	46	9	4
813	11	ζ	2	22	6	38	30	4
814	12	η	2	24	42	38	0 S.	4

Canis Major (*Vṛihat aśvapnmūrttiḥ*)

No.	No.	Bayer.	Longitude. S	°	′	Latitude. °	′	Mag.
815	1	α	3	10	27	39	30 S.	1
816	2	θ	3	13	3	34	45	4
817	3	μ	3	13	33	36	15	5
818	4	γ	3	16	33	38	0	4
819	5	ι	3	15	48	39	45	4
820	6	δ	3	11	33	43	0	5
821	7	ν^3	3	8	51	41	19	5
822	8	ν^1	3	8	39	42	30	5
823	9	β	3	3	33	41	30	3
824	10	ξ^1	3	7	12	46	36	4*
825	11	ξ^2	3	8	48	46	0	5
826	12	o^2	3	17	27	46	15	4
827	13	o^1	3	14	15	46	48	5
828	14	δ	3	19	18	48	21	3
829	15	ϵ	3	16	48	51	42	3
830	16	κ	3	14	33	55	15	4
831	17	ζ	3	4	15	53	45	3
832	18	η	3	25	33	50	45 S.	3

Informalae

No.	No.	Bayer.	Longitude. S	°	′	Latitude. °	′	Mag.
833	1		3	15	39	22	42 S.	4
834	2		2	29	9	60	45	4
835	3		3	3	3	58	45	4*
836	4		3	5	15	56	51	4
837	5		3	6	33	55	48	5

No.	No.	Bayer.	Longitude.			Latitude.		Mag.
			S	°	′	°	′	
838	6		2	20	48	55	21 S.	4
839	7		2	23	39	57	15	4
840	8		2	25	3	58	30	4
841	9		2	22	3	59	30	3
842	10		2	18	33	57	24	3
843	11		2	14	33	58	30 S.	4

Canis Minor (*Laghusvāna*)

No.	No.	Bayer.	Longitude.			Latitude.		Mag.
844	1	β	3	18	33	13	54 S.	4
845	2	α	3	22	30	16	0 S.	1

Argo Navis (*Saphina naukā*)

No.	No.	Bayer.	Longitude.			Latitude.		Mag.
846	1	e	4	3	24	42	42 S.	5
847	2	ι	4	7	18	43	33	3
848	3	ξ	4	2	21	45	12 S.	4
849	4	o	4	1	51	46	21	5
850	5	π	3	28	30	46	24	5
851	6	κ	3	29	18	47	42	4
852	7	ρ	3	29	3	49	9	4
853	8	τ	4	1	42	49	24	4
854	9	σ	4	1	27	49	6	5
855	10	χ	4	6	30*	49	48	4
856	11	υ	3	28	3	51	54	5
857	12	λ	3	27	9	58	30	3
858	13	*f*	4	3	15	55	30	5
859	14	φ¹	4	5	3	59	0	5
860	15	φ²	4	6	33	57	57	4
861	16	ψ²	4	9	51	58	9	4
862	17	δ	4	14	18	58	36	2
863	18	ω¹	4	10	51	60	0	5
864	19	ω²	4	14	48	59	51	5
865	20	A¹	4	14	3*	57	21	5
866	21	A²	4	15	33	57	49	5
867	22	p¹	4	29	24	52	30	4
868	23	p²	4	29	33	57	0	4
869	24	p³	4	28	3	59	0	4
870	25		5	2	33	60	15	4
871	26		5	2	27	61	24	4
872	27	o¹	4	22	54	21*	24	4
873	28	o²	4	22	42	49	6	4
874	29	o³	4	21	30	43	39	4
875	30	o⁴	4	23	3	43	15	4
876	31	ε	5	7	12	56	9	2
877	32		5	10	51	51	15	3
878	33	*i*	4	4	18	63	54	4
879	34	*r*	4	14	24	65	24	6
880	35	ζ	4	23	21	64	15	2
881	36	η	5	2	19	69	40	4
882	37		5	8	59	65	40	3
883	38	θ	5	15*	9*	65	50	3
884	39	ν	5	19	49	67*	20	3
885	40	*b*	5	24	49	62	50	4
886	41	c	6	1	49	62	15	4
887	42		2	26	21	67*	9	4
888	43	*g*	3	12	39	66	22*	3
889	44	α	3	10	58	75	0	1
890	45	*h*	3	22	49	71	45 S.	3

Hydra (*Sujā*)

No.	No.	Bayer.	Longitude.			Latitude.		Mag.
891	1	σ	4	7	36	14	33 S.	4
892	2	δ	4	6	33	12	30	4
893	3	ε	4	8	36	11	15 S.	4
894	4	η	4	10*	33	14	9	5
895	5	ζ	4	11	3	11	9	4
896	6	ω	4	13	48	12	9	6
897	7	θ	4	16	36	13	0	4
898	8	τ²	4	22	9	15	9	4
899	9	ι	4	23	36	14	39	4

No.	No.	Bayer.	Longitude.			Latitude.		Mag.
			S	°	'	°	'	
900	10	τ^1	4	22	3	16	42 S.	4
901	11	A	4	22	51	21	42	6
902	12	α	4	23	39	22	30	2
903	13	υ^1	5	2	18	26	0	4
904	14	υ^2	5	4	21	23	15	4
905	15	λ	5	5	18	22	0	4
906	16	μ	5	11	12	24	10*	3
907	17		4*	14	9	23	36	4
908	18		5	16	45	22	0	3
909	19	β	5	24	54	25	39	4
910	20		5	25	18	30	21	4
911	21		6	4	0	21*	42	4
912	22		6	7	18	33	48	4
913	23		6	9	18	31	15	3
914	24	γ	6	23	3	13	45	3
915	25	π	7	5	18	13	9 S.	3

Informatae

No.	No.	Bayer.	Longitude.			Latitude.		Mag.
916	1		4	6	24	22	39 S.	3
917	2		5	3	12	10	12	4

Crater (*Vāliyā vahu guṇa pātra*)

No.	No.	Bayer.	Longitude.			Latitude.		Mag.
918	1	α	5	20	3	22	42 S.	4
919	2	γ	5	26	3	19	45	4
920	3	δ	5	23	9	17	42	4
921	4	ζ	5	29	45	18	33	5
922	5	ϵ	5	22	30	13	21	4
923	6	η	6	2	3	17*	48*	4
924	7	θ	5	25	3	11	24 S.	4

Corvus

No.	No.	Bayer.	Longitude.			Latitude.		Mag.
			S	°	'	°	'	
925	1	α	6	8	21	22	0 S.	3
926	2	ϵ	6	8	6	19	15	3
927	3	ζ	6	10	33	18	15	5
928	4	γ	6	6	54	14	18 S.	3
929	5	δ	6	9	39	12	0	3
930	6	η	6	10	9	11	39	4
931	7	β	6	13	48	17	49	3

Centaurus (*Kaṁvūras*)

No.	No.	Bayer.	Longitude.			Latitude.		Mag.
932	1	*g*	7	4	33	22	9 S.	5
933	2	*h*	7	3	45	19	6	5
934	3	*i*	7	3	24	20	48 S.	4
935	4	*k*	7	4	9	20	0	5
936	5	ι	6	29	21	25	48	3
937	6	θ	7	8	48	21	57	3
938	7	ψ	7	2	33	27	45	5
939	8	*l*	7	11	33	23	0	4
940	9	o	7	12	42	24	0	4
941	10	π	7	15	24*	18	6	4
942	11	ρ	7	15	45	21	15	4
943	12	τ	7	6	52	28	45	4
944	13	υ	7	7	48	29	24	4
945	14	ϕ	7	9	3	27	45	4
946	15	*m*	7	10	15	26	42	4
947	16	κ	7	16	30	25	30*	3
948	17	σ	7	20	54	24	15	4
949	18	λ	7	11	3	32	48	3
950	19	*n*	7	10	51	30	0*	5
951	20	χ	7	10	3	30	48*	5
952	21	ω	7	6	3	34	54	5
953	22	o	7	2	9	37	42	5
954	23	μ	6	28	48	40	12	3
955	24	*c*	6	27	54	40	0	5
956	25	*p*	6	26	3	41	0	5
957	26	β	6	26	12	46	6	3
958	27	*e*	6	27	19	46	15	5
959	28		7	12	9	40	45	5

No.	No.	Bayer.	Longitude.			Latitude.		Mag.
			S	°	′	°	′	
960	29		7	10	9	43	0 S.	3
961	30			..		..		..
962	31	ν	7	3	49	51	10	2
963	32	ξ	7	9	9	51	40	2
964	33	f	7	0	9	55	10	3
965	34	ζ	7	4	59	55	20	2
966	35	α	8	2	9	41	10	1
967	36	γ	7	17	59	45	20	2
968	37	ε	7	8	29	49	10 S.	4

Lupus

No.	No.	Bayer.	Longitude.			Latitude.		Mag.
969	1	ο	7	21	15	25	0 S.	3
970	2	α	7	29	33	30	3	3
971	3	ζ	7	25	12	21	28*	4
972	4	η	7	26*	33	21	18	3
973	5	θ	7	26	45	25	12	4
974	6	π	7	23	27	27	30	5
975	7	β	7	24	15	29	12	5
976	8	ξ	7	26	39	29	0	5
977	9	ρ	7	26	12	29	57	5
978	10	ς	7	29	29*	33	10	4
979	11	τ	7	13	49	31	20 S.	5
980	12	ι	7	15	21	30	36	4
981	13	κ	7	16	33	29	24	5
982	14	ν	8	2	9	27*	18	4
983	15	μ	8	2	27	15	45	5
984	16	γ	7	29	3	13	21	5
985	17	λ	8	0	9	13	30	5
986	18	ε	7	20	41*	13	6	6
987	19	δ	7	21	6	11	30 S.	5

Ara

No.	No.	Bayer.	Longitude.			Latitude.		Mag.
988	1	γ	8	21	29	22	40 S.	6
989	2	ε	8	24	9	25	45	4

No.	No.	Bayer.	Longitude.			Latitude.		Mag.
			S	°	′	°	′	
990	3	δ	8	19	59	26	30 S.	4
991	4	α	8	14	20*	30	20	5
992	5	β	8	18	59	34	10	4
993	6	η	8	18	49	33	20	4
994	7	θ	8	14	39	34	0 S.	4

Corona Australis (*Mukuṭa*)

No.	No.	Bayer.	Longitude.			Latitude.		Mag.
995	1	α	9	2	15	22	0 S.	4
996	2	ε	9	5	42	21	18	6
997	3	ζ	9	6	24	20	30	6
998	4	β	9	8	0	19	51	5
999	5	η	9	9	24	18	18	5
1000	6	θ	9	10	18	17	18	5
1001	7	γ	9	10	9	16	12	5
1002	8	δ	9	9	42	15	15	5
1003	9	μ	9	8	24	15	12	5
1004	10	ν	9	8	9	14	39	6
1005	11	ι	9	5	33	15	0	5
1006	12	κ	9	3	15	16	0	5
1007	13	λ	9	2	9	18	36 S.	5

Piscis Austrinus (*Muchchhi yanûvî*)

No.	No.	Bayer.	Longitude.			Latitude.		Mag.
1008*	1	β	10	24	48	21	30 S.	4
1009	2	γ	10	28	18	23	30	4
1010	3	δ	10	29	3	23	48	4
1011	4	ε	10	28	54	17	45	4
1012	5	μ	10	20	3	21	0	5
1013	6	ζ	1(0)	26	15	16	45	5*
1014	7	λ	10	22	55	16	15	5
1015	8	η	10	19	30	15	30	5
1016	9	θ	10	15	27	16	54	5
1017	10	ι	10	14	33	18	33	4
1018	11	κ	10	14	33	23	15 S.	3

* Baily's No. 1008 is omitted.

No.	Longitudes.					
	Jai Singh.			Baily + 4° 8'.		
	S	°	'	°	'	'
1	2	24	23	2	24	27
10	3	19	57	3	19	51
28	4	15	18	4	15	48
29	4	47	15	4	17	15
49	9	19	58	9	19	18
53	0	11	9	0	1	9
84	10	11	9	0	11	9
94	6	29	25	6	29	24
111	7	8	38	7	8	42
115	7	10	26	7	10	36
118	7	14	39	7	15	39
132	8	6	15	8	6	12
150	9	16	18	9	15	18
205	1	8	54	2	8	54
219	2	26	30	2	6	30
221	2	28	51	2	18	51
225	2	25	9	2	15	9
232	8	19	12	8	19	21
257	8	26	44	8	26	45
275	8	21	8	8	21	12
350	1	5	39	1	5	19
360	1	1	55	1	0	21
361	1	0	21	1	1	15
378	1	20	28	1	20	18
405	1	4	33	2	4	33
411	2	18	13	2	13	18
439	3	27	21	2	27	21
449	4	3	42	4	3	43
462	2	26	6	4	17	15
465	2	24	27	4	24	27
471	4	21	30	4	20	30
497	5	29	27	5	24	27
526	7	12	15	7	12	0
530	7	17	20	7	17	24
545	7	28	45	7	28	48
562	8	20	19	8	20	39
564	8	24	0	8	24	3
568	9	1	36	9	1	6
590	9	12	21	9	12	51
593	9	28	33	9	18	33
603	10	1	57	10	1	27
634	11	3	27	11	3	21
637	11	7	30	11	7	3
646	11	21	42	11	2	42
649	11	6	52	11	6	12
665	11	16	12	11	6	12
666	11	6	23	11	7	3
669	11	23	20	11	23	18
676	11	29	24	11	19	24
723	1	13	3	0	13	3
743	2	11	3	2	21	3
752	2	8	24	2	8	21
753	2	8	47	2	8	42
764	2	29	39	2	18	39
772	2	7	21	2	7	51
797	1	20	30	1	20	33
803	1	11	48	0	11	48
810	2	15	48	2	15	51
855	4	6	30	4	6	3
865	4	14	3	4	14	7
883	5	15	9	5	14	29
894	4	10	33	4	8	33
907	4	14	9	5	14	9
941	7	15	24	7	15	14
972	7	26	33	7	27	33
978	7	29	29	7	29	49
986	7	20	41	7	20	48
991	8	14	20	8	14	29

A 1·2 Differences between Jai Singh's and Baily's versions of Ulugh Beg's catalogue.

Latitudes.

No.	Jai Singh.		Baily.	
	°	'	°	'
42	20	15	29	15
51	80	30	80	33
57	80	36	80	30
59	80	55	80	15
119	27	9	37	9
150	60	15	60	45
164	69	42	69	52
173	64	27	64	24
175	50	2	50	12
236	82	0	32	0
245		N.		S.
246		N.		S.
248		N.		S.
278	26	56	26	54
288	31	9	31	0
305	33	54	32	55
306	32	12	31	21
307	31	21	32	12
320	39	9	34	9
334	26	5	26	54
337	32	21	32	30
351	26	36	27	36
356	36	6	16	6
360	7	51	6	36
361	6	36	7	51
363	5	16	5	36
371		N.		S.
387	12	35	12	15
400		N.		S.
411	1	35	1	15
414	6	34	6	54
429	3	45	5	45
430	1	15	1	51
464	9	45	9	0
479	15	45	16	45
530		N.		S.
538	21	0	20	0
574		S.		N.
612	5	12	6	12
648		S.		N.
683	6	0	0	0
710	12	11	12	51
733	17	45	17	15
771	28	12	29	54
772	27	28	27	48
774	24	24	25	24
788	39	45	38	45
792	12	30	42	30
872	21	24	51	24
884	67	20	66	20
888	66	22	66	12
906	24	10	24	45
911	21	42	31	42
923	17	48	16	18
947	25	30	25	33
982	27	18	17	18

Star Magnitudes.

No.	J. S.	B.
8	4	3
70	4–3	3
128	3	4
246	5	4
496	5	4
619	3	4
620	..	3
824	4	5
835	4	5
1013	5	6

Additions to tables of differences.

No.	Jai Singh.	Baily.
	Longitude.	
	S ° '	S ° '
127	8 26 53	8 26 33
	Latitude.	
	° '	° '
555	15 55	15 15
571	2 28	2 18
887	67 9	66 9
950	30 0	30 48
951	30 48	30 0
971	21 28	21 18

A 2. MAHENDRA'S STAR LIST.

No.	Mahendra's List.			Ptolemy's values.		Differences.		Modern names and Magnitudes.	No. in Jai Singh.
	Name.	Longitude.	Latitude.	Longitude.	Latitude.	ΔLong	ΔLat		
		S ° ′	° ′	S ° ′	° ′	° ′			
1	..	0 6 43	+27 0	11 17 50	+27 0	18 53	0	δ Pegasi = α And. 2·1	313
2	Nadyantak .	0 19 43	—53 30	0 0 10	—53 30	19 30[1]	0	θ Eridani . .	802
3	..	0 22 43	+26 20	0 3 50	+26 20	18 53	0	β Andromedæ . 2·4	357
4	..	0 25 21	+ 7 20	0 6 40	+7 20	18 41[*]	0	γ Arietis . . 4·7	360
5	..	0 26 43	+51 20	0 7 50	+51 20	18 53	0	β Cassiopeia . 2·4	187
6	..	1 18 33	+23 0	0 29 40	+23 0	18 53	0	β Persei . . 2	201
7	..	1 23 43	+30 0	1 4 50	+30 0	18 53	0	α Persei . . 1·9	196
8	Brāhma .	2 1 33	—5 10	1 12 40	—5 10	18 53	0	α Tauri . . 1·1	391
9	At foot of twins.	2 8 43	—31 30	1 19 50	—31 30	18 53	0	β Orionis . 0·3	765
10	Its left shoulder.	2 12 53	—17 30	1 24 0	—17 30	18 53	0	γ Orionis . 1·7	733
11	Shaḍasya .	2 13 53	—22 30	1 25 0	—22 30	18 53	0	α Aurigæ . . 0·2	221
12	Ārdra . .	2 20 53	—17 0	2 2 0	—17 0	18 53	0	α Orionis . . 0·9	732
13	Agastya .	3 6 4	—75 0	2 17 0	—75 0	18 54[*]	0	α Argus . . —0·8	889
14	..	3 6 33	—39 10	2 17 40	—39 0	18 53	0	α Canis Majoris— 1·6	815
15	..	3 12 33	+9 40	2 23 29	+9 40	19 4[*]	0	α Geminorum . 2	421
16	Vyādhāniya .	3 18 43	—16 10	2 29 10	—16 10	18 33[2]	0	α Canis Minoris 0·5	845
17	Maghā .	4 21 23	+0 10	4 2 30	+0 10	18 53	0	α Leonis . . 1·3	466
18	..	5 13 23	+11 50	4 24 30	+11 50	18 53	0	β Leonis . . 2·2	485
19	..	5 27 33	—14 50	5 13 30	—14 50	14 3[*]	0	γ Corvi . . 2·8	928
20	Chitrā . .	6 15 33	—2 0	5 26 40	—2 0	18 53	0	α Virginis . 1·2	507
21	Svāti . .	6 15 53	+31 30	5 27 0	+31 30	18 53	0	α Bootis . . 0·2	110
22	Viśākhā .	7 3 33	+44 30	6 14 40	+44 30	18 53	0	α Coronæ Borealis 2·3	111
23	Jyeshṭhā .	8 1 33	—4 0	7 12 40	—4 0	18 53	0	α Scorpii . . 1·2	550
24	At the edge of Dhanus.	8 13 43	+36 0	7 24 50	+36 0	18 53	0	α Ophiuci . 2·1	232
25	Mūla . .	8 20 3	—13 10	8 1 10	—13 15	18 53	0° 5′	γ Telescopii . *Neb.*	564
26	..	8 25 34	+2 50	8 6 40	+2 50	18 54[*]	0	μ Sagitarii . 3·8	571
27	Abhijit .	9 6 13	+62 0	8 17 20	+62 0	18 53	0	α Lyræ . . 0·14	148
28	..	9 22 43	+29 10	9 3 50	+29 10	18 53	0	α Aquilæ . . 0·9	286
29	..	10 25 53	—23 0	10 7 0	—23 0	18 53	0[3]	α Pisc. Aust. . 1·3	667
30	..	10 28 3	+60 0	10 9 10	+60 0	18 53	0	α Cygni . . 1·3	162
31	..	11 21 3	+31 0	11 2 10	+31 0	18 53	0	β Pegasi . . 2·6	316
32	..	11 23 13	—9 40	11 4 40	—9 40	18 53	0	ι Ceti . . 3·7	729

[1] See the interesting note in Peters and Knobel's *Ptol. Cat.*, p. 110. The readings for longitude vary.
[2] The readings vary. [3] Some authorities give —20° 20, but Baily gives —23° 0. See *Ptol. Cat.* p. 113.

A 3. THE SŪRYA SIDDHĀNTA STAR LIST.

Hindu Names.	Probable identification of stars.	Sūrya Siddhānta				Flamsteed's values reduced to A.D. 560.		Nakshatra of which the star is an indicator.
		Polar Long.	Polar Lat.	Reduced				
				Long.	Lat.	Long.	Lat.	
		° ′	°	° ′	° ′	° ′	° ′	
	β Arietis . .	8 0	10 N.	11 59	9 11 N.	13 56	8 28 N.	1. Aśvinī.
	35 Arietis . .	20 0	12 N.	24 35	11 6 N.	26 54	11 17 N.	2. Bharaṇī.
	η Tauri . .	37 30	5 N.	39 8	4 44 N.	39 58	4 1 N.	3. Kṛittikā.
	α Tauri . .	49 30	5 S.	48 9	4 49 S.	49 45	5 30 S.	4. Rohiṇī.
Brahmahṛidaya.	α Aurigæ . .	..	..	60 29	28 53 N.	61 50	21 52 N.	
Agni or Hutabhuj.	β Tauri . .	..	..	54 5	7 44 N.	62 32	5 22 N.	
	λ Orionis . .	63 0	10 S.	61 3	9 49 S.	63 40	13 25 S.	5. Mṛigaśiras.
	α Orionis . .	67 20	9 S.	65 50	8 53 S.	68 43	16 4 S.	6. Ārdrā.
Prajāpati or Brahma.	δ Aurigæ . .	..	..	67 11	36 49 N.	69 54	30 49 N.	
Mṛigavyādha or Lubdhaka.	α Canis Majoris	..	..	76 23	39 52 S.	84 7	39 32 S.	
Agastya .	α Navis . .	..	..	90 0	80 0 S.	85 4	75 50 S.	
	β Geminorum .	93 0	6 N.	92 55	6 0 N.	93 14	6 39 S.	7. Punarvasu.
	δ Cancri . .	106 0	0 0	106 0	0 0	108 42	0 4 N.	8. Pushya.
	ε Hydræ .	109 0	7 S.	109 59	6 56 S.	112 20	11 8 S.	9. Āśleshā.
	α Leonis . .	129 0	0 0	129 0	0 0	129 49	0 27 N.	10. Maghā.
	δ Leonis . .	144 0	12 N.	139 58	11 19 N.	141 15	14 19 N.	11. P. Phālgunī.
	β Leonis . .	155 0	13 N.	150 10	12 5 N.	151 37	12 17 N.	12. U. Phālgunī.
Āpas . .	δ Virginis .	..	..	176 23	8 15 N.	171 38	8 38 N.	
	δ Corvi . .	170 0	11 S.	174 22	10 6 S.	173 27	12 10 S.	13. Hasta.
Apāmvatsa .	θ Virginis .	..	..	178 48	2 45 N.	178 12	1 45 N.	
	α Virginis .	180 0	2 S.	180 48	1 50 S.	183 49	2 2 S.	14. Chitrā.
	α Bootis . .	199 0	37 N.	183 2	33 50 N.	184 12	30 57 N.	15. Svātī.
	ι Libræ . .	213 0	1½ S.	213 31	1 25 S.	211 0	1 48 S.	16. Viśākhā.
	δ Scorpii . .	224 0	3 S.	224 44	2 52 S.	222 34	1 57 S	17. Anurādhā.
	α Scorpii . .	229 0	4 S.	230 7	3 50 S.	229 44	4 31 S	18. Jyeshṭhā.
	λ Scorpii . .	241 0	9 S.	242 52	8 48 S.	244 33	13 44 S	19. Mūla.
	δ Sagittarii .	254 0	5½ S.	254 39	5 28 S.	251 32	6 25 S	20. P. Ashāḍhā.
	σ Sagittarii .	260 0	5 S.	260 23	4 59 S.	262 21	3 24 S	21. U. Ashāḍhā.
	α Lyræ . .	266 40	60 N.	264 10	59 58 N.	265 15	61 46 N.	22. Abhijit.
	α Aquilæ . .	280 0	30 N.	282 29	29 54 N.	281 41	29 19 N.	23. Śravaṇa.
	β Dolphini .	290 0	36 N.	296 5	35 33 N.	296 19	31 57 N.	24. Śravishṭha.
	λ Aquarii .	320 0	½ S.	319 50	0 28 S.	321 33	0 23 S.	25. Śatabhishaj.
	α Pegasi . .	326 0	24 N.	334 25	22 30 N.	333 27	19 25 N.	26. P. Bhādrapadā.
	γ Pegasi . .	537 0	26 N.	347 16	24 1 N.	349 8	25 41 N.	27. U. Bhādrapadā.
	ζ Piscium .	359 0	0 0	359 50	0 0	359 50	0 0	28. Revatī.

The Zarqālī Instrument.

Names of Stars.	Approximate position on the instrument.				Position according to Flamsteed.			
	R. A.	Declin.	Long.	Lat.	R. A.	Declin.	Long.	Lat.
	°	°	S °		° ′	° ′	S ° ′	° ′
Dhanab al-Qiṭus (β Ceti) .	6¼	—20	11 27½	—22	7 0	—19 41	11 28 13	20 24
Sadr al-Qiṭus (ι Ceti) .	38	—16	0 29½	—29	36 9½	—13 13	0 29 0	—26 0
Fam al-Qiṭus (α Ceti, *Menkar*).	37½	+2	1 6	—13	41 32	+2 50	1 9 59	—12 37
Rijl al-juzā al-Yasri (β Orionis, *Rigel*).	75	—9¼	2 12½	—31½	74 55	—8 36	2 12 30	—31 10
Farad al-Shaja' (α Hydræ).	137½	—7¾	4 22½	—22½	138 5	—7 20	4 22 58	—22 25
Rās al-Āsad (μ Leonis) .	142	+28	4 15½	+12	143 45	27 27	4 17 6	+12 9
Qā'da Baṭih (α Crateris) .	160	—16½	5 19	—22¾	161 12	—16 39	5 19 27	—22 42
Ẓahr al-Āsad (α Leonis) .	162	+22¾	5 5½	+14	164 23	22 12	5 6 57	+14 9
Simāk Rāmiḥ (α Bootis *Arcturus*).	210	+22	6 20	+32	210 22½	+20 49	6 19 54	+30 57
Nasr-Wāqi' (α Lyræ, *Vega*)	277½	+38½	9 10½	62 0	276 36	+38 32	9 10 57	+61 45½
Dhanab al-dajājah (α Cygni).	308	+44	11 1½	+60½	307 40½	+44 12	11 1 1½	+59 57
Dhanab al-Jadi (ζ Capricorni).	321½	—17	10 18	—2	322 29	—17 29	10 19 13	—2 32
Fam al-Faras (ι Pegasi) .	322	+8½	10 27	+22	322 14	+8 28	10 27 34	+22 7
Mankib, al-Faras (β Pegasi)	342	+27	11 24½	+31 0	342 11½	+26 24	11 25 2	+31 8

[1] Flamsteed had been at work at Greenwich for four years when the Zarqālī instrument was made. The instrument is dated A.D. 1680 and Flamsteed's catalogue is for 1689.

APPENDIX B.

Astrological Tables.

B.—ASTROLOGY.

The instruments, both brass and masonry, were sometimes used for astrological purposes and in some cases were graduated specifically for such purposes. The following notes and the accompanying tables are concerned only with such astrological matters[1] as are exhibited by the instruments, and relate to :—

(1) The ascendant or rising sign.

(2) Houses (*Domus Cœli*).

(3) Trigon or Triplicities and their Regents.

(4) Terms or Limits.

(5) Decans and Faces.

(6) Planetary Domiciles.

(7) Septenaries, Nonenaries, Duodenaries.

The **Ascendant** or 'horoscope' is the point of the ecliptic rising to the horizon at the given moment. Its determination is the first and most important astrological problem. By means of the Jai Prakāś (page 37) or the Kapāla the ascendant could be determined by inspection. On the Jai Prakāś the shadow of the intersection of the cross wires shows not only the position of the sun but also the sign that is on the meridian, from which the rising sign could be deduced ; while the Kapāla shows the rising sign itself.

On the astrolabe the position of the sun on the ecliptic (its longitude) and its position for the day (its altitude) being known, the only operation required is to turn the 'ankabūt (*aranea*), so that the part of the ecliptic in which the sun is, lies on the proper altitude circle (almucantarat), and then the rising sign or the point of the ecliptic on the horizon can be at once read off.

On the 'Jaipur B' astrolabe is given a table of the times of rising of the signs (see p. 23) from which the ascendant, etc., could be calculated, if the position of the sun were known, for any of the given latitudes.

The ecliptic was divided into 12 equal divisions or signs of the zodiac which, owing to the obliquity, took different times to rise and set. This problem of ascensions ('Αναφοραί) became of great importance because it affected the position of the four 'centres.' (1) The 'horoscope' or 'ascendant.' (2) Superior culmination. (3) The descendant. (4) Inferior culmination. These, in consequence of the variable time of the risings and settings of the signs, are not at intervals of right angles as the early 'Egyptian' astrologers assumed.

This problem of the 'anaphorai, is most interesting historically. Hypsicles and Hipparchus (second century B.C.) studied it and Ptolemy gave the correct solution. Paulus of Alexandria (third century of our era) animadverted on the erroneous methods[2] employed and exhibited the 'anaphorai according to Ptolemy.

---

[1] We are not here concerned with the fact that the fundamental assumptions in connexion with astrology are false. Ptolemy assumed influences emanating from celestial bodies which tended to make the nature of the subject affected similar to the agent. The Arabs considered the heavenly bodies rather as indicators than agents. Neither of these assumptions is warranted by any combination of observation and reasoning ; both were the result of false ideas that have been long since discarded except by the un-learned.

[2] *E.g.*, by Manilius. See Delambre, Vol. 1, p. 253.

The following incorrect table[1] from Petosiris for the first and second climates may be compared with that given on page 23 :—

	Rising.		Setting.		
	1st Climate.	2nd Climate.	1st Climate.	2nd Climate.	
	° ′ h. m. s.	° h. m.	° ′	°	
Aries	21 40 = 1 26 40	20 = 1 20	38 20	40	Pisces.
Taurus	25 0 = 1 40 0	24 = 1 36	35 0	36	Aquarius.
Gemini	28 20 = 1 53 20	28 = 1 52	31 40	32	Capricornus.
Cancer	31 40 = 2 6 40	32 = 2 8	28 20	28	Sagittarius.
Leo	35 0 = 2 20 0	36 = 2 24	25 0	24	Scorpio.
Virgo	38 20 = 2 33 20	40 = 2 40	21 40	20	Libra.
Total .	180 0 = 12 0 0	180 = 12 0	180 0	180	

The table for the second climate is reproduced in the *Brihaj jātaka*[2] of Varāha Mihira (c. A.D. 550) and is therefore of considerable historical interest. Varāha Mihira lived at Ujjain, which the Hindus placed at the middle of the second climate.[3] The *Sūrya Siddhānta* (iii, 42 *f.*) gives the correct rule for determining the times of rising (*Udayāsavas*) of the signs, but gives 24° as the obliquity.

Houses.—The astrological houses (*Domus cœli*) must not be confused with the planetary domiciles. The system of twelve houses was not altogether accepted by Ptolemy but since the date of Sextus Empiricus (3rd century of the present era) it has been in universal use.[4] In plate V and figure 17 the twelve houses are shown. The boundary lines pass through the intersection (*H*) of the horizon and meridian and cut the equator at equal intervals of thirty degrees. The points at which these boundary lines cut the ecliptic are termed the cusps (*Cuspides domorum*) and four of these are at once seen (*E*, t_s, *W* and *A* in plate V) but to find the others is a mathematical problem of some little difficulty and occupied the attention of al-Battānī, Regiomontanus, Jean-Antoine Magini (1556-1617) and others. According to Delambre (p. 501) the Arabs divide the south-east quadrant of the equator (*S.E.* in plate *V*) into spaces each equivalent to two temporal day hours and the next quadrant (*E.N.*) into corresponding spaces of night hours.[5] Campanus and Gazulus divide the prime vertical (*EZW*) instead of the equator into equal divisions, and so on.

Trigons or Triplicities.—The Greek astrologers tried almost every possible combination of the signs of zodiac. The signs being represented by equal spaces on the circumference of the circle, pairs were associated by parallel lines (*Antiscia* of Firmicus, etc.), by diameters, by squares, by hexagons, and by triangles, inscribed in the circle.

The triangular aspect was considered the most beneficial and Ptolemy gives as the reason for this that the trigon unites signs of the same sex,[6] but Bouché-Leclercq suggests as the motif the part that the triad played in oriental religions.

[1] From Bouché-Leclercq's *Astrologie Grecque*, p. 269.

[2] The rule reads : " The numbers five, six, seven, eight, nine and ten each multiplied by four are respectively the measurements of the first six signs from Mesha (Aries) ; and these reversed become the *lagna mânas* of the last six signs." (I, 1f.)

[3] The Hindus placed Ujjain on the tropic of Cancer, *i.e.*, latitude 23° 51′ 15″ according to Ptolemy.

[4] Manilius (first century B.C.) had employed a scheme of eight houses.

[5] But see figure 17 which does not support this practice.

[6] The sex of the signs is determined by the Pythagorean view of numbers, which shows the odd numbers as masculine and the even as feminine. The signs are alternately masculine and feminine starting with Aries which is masculine *L'Astrologie Grecque*, p. 154.

The trigons or triplicities are sets of three signs that are 120 degrees apart. The trigons in order are—

(a)	(b)	(c)	
(i) ARIES	LEO	SAGITTARIUS	Masculine, Royal Trigon.
(ii) TAURUS	VIRGO	CAPRICORNUS	Feminine.
(iii) GEMINI	LIBRA	AQUARIUS	Masculine, Human Trigon.
(iv) CANCER	SCORPIO	PISCES	Feminine.

The **Lords or Regents** of the trigons are given on several astrolabes (p. 23). The origin of the arrangement is obscure but according to Geminus the orientation is determined by the direction of the wind when the moon occupies one of the signs of the trigons, etc., etc.

The regents[1] are—

(a)	(b)	(c)
(i) Saturn	Sun	Jupiter.
(ii) Venus	Moon	Mars.
(iii) Mercury	Saturn	Jupiter.
(iv) Moon	Mars	Venus.

The tables on the astrolabes correspond to these principles but add other information. The table of trigons on 'Jaipur A' is given on p. 23; the corresponding tables on 'Herāt C' is seen in figure 12. In both of these tables the trigons are classified in order as (i) Fiery, (ii) Earthy, (iii) Airy and (iv) Watery.[2]

Terms or Limits (ὅρια).—Fractions of each sign of 30 degrees are distributed among the five planets and the amount allotted to each planet determines the quantity of its influence. The planets are arranged in an order which varies for each sign; but the order never forms an intelligible series and the determining causes of the scheme are not understood.[3] The system followed on all the astrolabes examined is that known to Dorothea of Sidon, Firmicus and Paulus Alexandrinus.[4] It did not satisfy Ptolemy who tried to introduce a rational order but failed for the Egyptian system is the only one generally recognised.

It is curious that some astrologers tried to explain the numbers as the times of rising of the respective planets but Ptolemy pointed out that these depend upon the latitude of the observer, etc.; Demophilus said the numbers represented the periods of revolution of the planets; some say the number allotted to each planet represents the number of years of life that it can impart to the individual born under its influence, etc.

[1] According to Dorothea of Sidon Ptolemy gives Mars for Saturn in i, (a).

[2] According to Alberuni (ii, 220) the Hindus do not refer the *aspectus trigoni* to the elements.

[3] The *Bṛihāj-jātaka* (i, 7) seems to aim at a systematic arrangement but the text is not at all clear.

[4] Bouché-Leclercq, p. 206f.

The question of ὅρια made discord with the astrologers. "Apollinaris," writes Demophilus, "disagrees with Ptolemy about the distribution of the ὅρια and they both with Thrasyllus, Petosiris, and other ancient authorities."[1] The complete table of terms is given below (p. 124).

Decans and Faces.—The decans (δεκανοί) are parts of the zodiac each equal to 10 degrees. Each division of time had its protecting genius or chronocrator and the 36 decans are possibly of such or religious origin and correspond to 36 protecting divinities. Hermes Trismegistus speaks of the 36 decans as 'vigilant guardians, inspectors of the universe.'

The system of decans imposes three kinds of influence: (1) that of the decans themselves, (2) that of the stars that rise at the same time and (3) that of the 'figures' or 'faces' (πρόσωπα).

The order of the decans, after Firmicus and Paulus Alexandrinus, etc., and on the astrolabes is as follows:—

Signs.	Decans.		
ARIES	Mars	Sun	Venus.
TAURUS	Mercury	Moon	Saturn.
GEMINI	Jupiter	Mars	Sun.
CANCER	Venus	Mercury	Moon.
LEO	Saturn	Jupiter	Mars.
VIRGO	Sun	Venus	Mercury.
LIBRA	Moon	Saturn	Jupiter.
SCORPIO	Mars	Sun	Venus.
SAGITTARIUS	Mercury	Moon	Saturn.
CAPRICORNUS	Jupiter	Mars	Sun.
AQUARIUS	Venus	Mercury	Moon.
PISCES	Saturn	Jupiter	Mars.

It will be noticed that the decans read vertically in the above table are in the order of the days of the week.

Planetary Houses.—The idea that the planets, as divinities rejoiced in particular positions seems to have come from the East (Babylon ?). The scheme seems to have been evolved by equating the planets in order with the signs starting from the beginning of the calendar year, thus:—

Houses.	Planets.	Houses.
AQUARIUS	1 Saturn	CAPRICORNUS.
PISCES	2 Jupiter	SAGITTARIUS.
ARIES	3 Mars	SCORPIO.
TAURUS	4 Venus	LIBRA.
GEMINI	5 Mercury	VIRGO.
LEO	6 Moon—Sun 6'	CANCER.

[1] Bouché-Leclercq, p. 223.

On the astrolabes (B. and E.) the arrangement is as follows :—

AQUARIUS	1 Saturn	5 Mercury	4 Venus.
PISCES	2 Jupiter	6 Moon	3 Mars.
ARIES	3 Mars	6′ Sun	2 Jupiter.
TAURUS	4 Venus	5 Mercury	1 Saturn.
GEMINI	5 Mercury	4 Venus	1 Saturn.
CANCER	6 Moon	3 Mars	2 Jupiter.
LEO	6 Sun	2 Jupiter	3 Mars.
VIRGO	5 Mercury	1 Saturn	4 Venus.
LIBRA	4 Venus	1 Saturn	5 Mercury.
SCORPIO	3 Mars	2 Jupiter	6 Moon.
SAGITTARIUS	2 Jupiter	3 Mars	6′ Sun.
CAPRICORNUS	1 Saturn	4 Venus	5 Venus.

The *Brihaj Jātaka* (i, 6) says : "Mars, Venus, Mercury, the Moon, the Sun, Mercury, Venus, Mars, Jupiter, Saturn, Saturn, Jupiter, are successively the rulers of the twelve houses, Meśá, Vṛisha, Mithuna, etc., as well as of the Navāṁśas and Dvādaśāṁśas of the houses."

Septenaries.—These occur only on the Zarqālī instrument and are as follows :—

AQUARIUS	1 Saturn	2 Jupiter	3 Mars	4 Sun	5 Venus	6 Mercury	7 Moon	CAPRICORNUS.
PISCES	2 Jupiter	3 Mars	4 Sun	5 Venus	6 Mercury	7 Moon	1 Saturn	SAGITTARIUS.
ARIES	3 Mars	4 Sun	5 Venus	6 Mercury	7 Moon	1 Saturn	2 Jupiter	SCORPIO.
TAURUS	5 Venus	6 Mercury	7 Moon	1 Saturn	2 Jupiter	2 Mars	4 Sun	LIBRA.
GEMINI	6 Mercury	7 Moon	1 Saturn	2 Jupiter	3 Mars	4 Sun	5 Venus	VIRGO.
CANCER	7 Moon	1 Saturn	2 Jupiter	3 Mars	4 Sun	5 Venus	6 Mercury	
	4 Sun	5 Venus	6 Mercury	7 Moon	1 Saturn	2 Jupiter	3 Mars	LEO.

The horizontal order starting with Aquarius is the standard order[1] of the planets ; the vertical order (of the first column) is the order of the planetary domiciles (p. 122).

Novenaries.—If the planets be arranged in the domiciliary order, (1) Saturn, (2) Jupiter, (3) Mars, (4) Venus, (5) Mercury, (6) Moon, (6′) Sun, in nines for each trigon starting with Mars we get the following arrangement :—

TRIGONS.	1	2	3	4	5	6	7	8	9
1. ARIES, LEO, SAGITTARIUS.	3 Mars	4 Venus	5 Mercury	6 Moon	6′ Sun	5 Mercury	4 Venus	3 Mars	2 Jupiter.
2. TAURUS, VIRGO, CAPRICORNUS.	1 Saturn	1 Saturn	2 Jupiter	3 Mars	4 Venus	5 Mercury	6 Moon	6′ Sun	5 Mercury.
3. GEMINI, LIBRA, AQUARIUS.	4 Venus	3 Mars	2 Jupiter	1 Saturn	1 Saturn	2 Jupiter	3 Mars	4 Venus	5 Mercury.
4. CANCER, SCORPIO, PISCES.	6′ Moon	6′ Sun	5 Mercury	4 Venus	3 Mars	2 Jupiter	1 Saturn	1 Saturn	2 Jupiter.

and this is really the arrangement given on the Zarqālī instrument and it is implied in the *Bṛihāj jātaka* (i, 6) which says : "The rulers of the nine Navāṁsas of Mesha, Makara, Tulā, Karkaṭa are the same who rule the nine houses beginning with Mesha."

[1] The classical (Plato, Aristotle, Eratosthenes, etc.) order was, *Moon, Sun* (*Venus, Mercury*), *Mars, Jupiter, Saturn.* Later *Mercury* and *Venus* were interchanged (Heraclitus) and about the time of Hipparchus the order was made *Moon, Mercury, Venus, Sun, Mars, Jupiter, Saturn.*

Duodenaries.—The Zarqālī instrument (E.) gives the planets also in twelves for each sign (figure 19). The arrangement is exhibited below : the order is the domiciliary order both horizontally and vertically.

	1	2	3	4	5	6	7	8	9	10	11	12
AQUARIUS	Saturn	Jupiter	Mars	Venus	Mercury	Moon	Sun	Mercury	Venus	Mars	Jupiter	Saturn.
PISCES	Jupiter	Mars	Venus	Mercury	Moon	Sun	Mercury	Venus	Mars	Jupiter	Saturn	Saturn.
ARIES	Mars	Venus	Mercury	Moon	Sun	Mercury	Venus	Mars	Jupiter	Saturn	Saturn	Jupiter.
TAURUS	Venus	Mercury	Moon	Sun	Mercury	Venus	Mars	Jupiter	Saturn	Saturn	Jupiter	Mars.
GEMINI	Mercury	Moon	Sun	Mercury	Venus	Mars	Jupiter	Saturn	Saturn	Jupiter	Mars	Venus.
CANCER	Moon	Sun	Mercury	Venus	Mars	Jupiter	Saturn	Saturn	Jupiter	Mars	Venus	Mercury.
LEO	Sun	Mercury	Venus	Mars	Jupiter	Saturn	Saturn	Jupiter	Mars	Venus	Mercury	Moon.
VIRGO	Mercury	Venus	Mars	Jupiter	Saturn	Saturn	Jupiter	Mars	Venus	Mercury	Moon	Sun.
LIBRA	Venus	Mars	Jupiter	Saturn	Saturn	Jupiter	Mars	Venus	Mercury	Moon	Sun	Mercury.
SCORPIO	Mars	Jupiter	Saturn	Saturn	Jupiter	Mars	Venus	Mercury	Moon	Sun	Mercury	Venus.
SAGITTARIUS	Jupiter	Saturn	Saturn	Jupiter	Mars	Venus	Mercury	Moon	Sun	Mercury	Venus	Mars.
CAPRICORNUS	Saturn	Saturn	Jupiter	Mars	Venus	Mercury	Moon	Sun	Mercury	Venus	Mars	Jupiter.

SUMMARY OF THE ASTROLOGICAL TABLES THAT OCCUR ON THE INSTRUMENTS EXAMINED.

Signs.			Terms or Limits.					Number of Trigon.	Trigons.		
									Day Regents.	Night Regents.	Nature of Trigon.
0	♈	Aries	Jupiter 6	Venus 6	Mercury 8	Mars 5	Saturn 5	i	Sun Jupiter Saturn	Jupiter Sun Saturn	Royal, Fiery
1	♉	Taurus	Venus 8	Mercury 6	Jupiter 8*	Saturn 5*	Mars 3	ii	Venus Moon Mars	Moon Venus Mars	Earthy
2	♊	Gemini	Mercury 6	Jupiter 8	Venus 5	Mars 7	Saturn 6	iii	Saturn Mercury Jupiter	Mercury Saturn Jupiter	Human, Airy
3	♋	Cancer	Mars 7	Venus 0	Mercury 6	Jupiter 7	Saturn 4	iv	Venus Mars Moon	Mars Venus Moon	Watery
4	♌	Leo	Jupiter 6	Venus 5	Saturn 7	Mercury 6	Mars 6	i	Sun Jupiter Saturn	Jupiter Sun Saturn	Royal, Fiery
5	♍	Virgo	Mercury 7	Venus 10	Jupiter 4	Mars 7	Saturn 2	ii	Venus Moon Mars	Moon Venus Mars	Earthy
6	♎	Libra	Saturn 6	Mercury 8	Jupiter 7	*Venus 7	Mars 2	iii	Saturn Mercury Jupiter	Mercury Saturn Jupiter	Human, Air
7	♏	Scorpio	Mars 7	Venus 4	Mercury 8	Jupiter 5	Saturn 6	iv	Venus Mars Moon	Mars Venus Moon	Watery
8	♐	Sagittarius	Jupiter 12	Venus 5	Mercury 4	Saturn 5	Mars 4	i	Sun Jupiter Saturn	Jupiter Sun Saturn	Royal, Fiery
9	♑	Capricornus	Mercury 7	Jupiter 7	Venus 8	Saturn 4	Mars 4	ii	Venus Moon Mars	Moon Venus Mars	Earthy
10	♒	Aquarius	Mercury 7	Venus 6	Jupiter 7	Mars 5	Saturn 5	iii	Saturn Mercury Jupiter	Mercury Saturn Jupiter	Human, Airy
11	♓	Pisces	Venus 12	Jupiter 4	Mercury 3	Mars 9	Saturn 2	iv	Venus Mars Moon	Mars Venus Moon	Watery

SUMMARY OF THE ASTROLOGICAL TABLES THAT OCCUR ON THE INSTRUMENTS EXAMINED—*contd.*

Faces or Decans.			Planets of which the sign is the domicile.			Septenaries.			Novenaries		Duodenaries		Mansions of the moon.	Signs.		
10	20	30	10	20	30											
Mars	Sun	Venus	Mars	Sun	Jupiter	Mars	Sun	Venus, etc.	Mars	Venus, etc.	Mars	Venus, etc.	Sharatin. Butin.	Aries	♈	0
Mercury	Moon	Saturn	Venus	Mercury	Saturn	Venus	Mercury	Moon, etc.	Saturn	Saturn, etc.	Venus	Mercury, etc.	Thuraiya. Dabaran. Haq'ah.	Taurus	♉	1
Jupiter	Mars	Sun	Mercury	Venus	Saturn	Mercury	Moon	Saturn, etc.	Venus	Mars, etc.	Mercury	Moon, etc.	Han'ah. Dhira. Nathrah.	Gemini	♊	2
Venus	Mercury	Moon	Moon	Mars	Jupiter	Moon	Saturn	Jupiter, etc	Moon	Sun, etc.	Moon	Sun, etc.	Tarf. Jabhah.	Cancer	♋	3
Saturn	Jupiter	Mars	Sun	Jupiter	Mars	Sun	Venus	Mercury, etc.	Mars	Venus, etc.	Sun	Mercury, etc.	Zubrah. Sarfah.	Leo	♌	4
Sun	Venus	Mercury	Mercury	Saturn	Venus	Mercury	Moon	Saturn, etc.	Saturn	Saturn, etc.	Mercury	Venus, etc.	'Auwa. Simak. Ghafr.	Virgo	♍	5
Moon	Saturn	Jupiter	Venus	Saturn	Mercury	Venus	Mercury	Moon, etc.	Venus	Mars, etc.	Venus	Mars, etc.	Zubanan. Iklil.	Libra	♎	6
Mars	Sun	Venus	Mars	Jupiter	Moon	Mars	Sun	Venus, etc.	Moon	Sun, etc.	Mars	Jupiter, etc.	Qalb. Shaulah. Na'aim.	Scorpio	♏	7
Mercury	Moon	Saturn	Jupiter	Mars	Sun	Jupiter	Mars	Sun, etc.	Mars	Venus, etc.	Jupiter	Saturn, etc.	Baldah. Sa'd al-Dhabih.	Sagittarius	♐	8
Jupiter	Mars	Sun	Saturn	Venus	Venus	Saturn	Jupiter	Mars, etc.	Saturn	Saturn, etc.	Saturn	Saturn, etc.	Sa'd Bula'. Sa'd al-Su'ud.	Capricornus	♑	9
Venus	Mercury	Moon	Saturn	Mercury	Venus	Saturn	Jupiter	Mars, etc.	Venus	Mars, etc.	Saturn	Jupiter, etc.	Sa'd al-Akhbiyah. al-Fargh al-Muqdim.	Aquarius	♒	10
Saturn	Jupiter	Mars	Jupiter	Moon	Mars	Jupiter	Mars	Sun, et	Moon	Sun, etc.	Jupiter	Mars, etc.	al-Fargh al-Mukhir. Batn al-Hut.	Pisces	♓	11
The vertical order is that of the days of the week.						*The order is horizontally the planetary but vertically for the first column of the domiciliary order.*			*The domiciliary order by Trigons is followed.*		*Both horizontally and vertically the order is the domiciliary.*					

APPENDIX C.

Geographical Elements.

C. (1) Astrolabe Gazetteer.[1]

Place.	Longitude.		Latitude.		Inhirāf.		Distance.
	°	′	°	′	°	′	
Mecca	77 *40*	10 *41*	21 *21*	40 *20*	0	0	0
Madina	75 *39*	20 *53*	25 *24*	0 *25*	34	10	86
Ardabīl	82 *48*	30 *19*	38 *38*	0 *12*	17	33	377
Astarābād	89 *54*	35 *26*	36 *36*	50 *46*	38	48	416
Baghdād	80 *44*	0 *25*	33 *33*	25 *21*	12	45	266
Baital Maqaddas[2]	66 *35*	30 *14*	31 *31*	50 *47*	45	16	309
Banāras	117[3] *83*	20 *0*	26 *25*	15 *18*	..		..
Balkh	101 *66*	0 *48*	36 *36*	41 *45*	60	36	..
Başra	84 *47*	30 *50*	30 *30*	0 *30*	34	19	230
Dāmghān	88 *54*	55 *19*	36 *36*	20 *16*	38	0	382
Damīaţ	63 *31*	30 *23*	31 *31*	25 *48*	..		..
Dilhi[4]	113 *77*	35 *0*	28 *28*	39 *44*	87	34	763
Dimishq	70 *36*	0 *18*	33 *33*	10 *30*	30	31	293
Golkandah	114 *78*	39 *26*	18 *7*	0 *12*	..		..
Gwaliār	115 *78*	0 *26*	26 *26*	10 *18*	..		..
Halb[5]	72 *37*	10 *0*	35 *36*	30 *10*	58	29	..
Hamadān	83 *48*	0 *20*	35 *34*	10 *50*	22	17	320
Herāt	94 *62*	20 *9*	34 *34*	30 *28*	53	17	439
Işfahān	86 *51*	40 *44*	32 *32*	25 *39*	40	28	330
Iskandariyah[6]	61 *29*	54 *51*	30 *31*	58 *11½*	..		..
Kābul	104 *69*	40 *18*	34 *34*	7 *30*	69	57	604
Kāshān	86	0	34	0	..		..
Kashmir	108	0	35	0	68	44	605
Kufah	79	30	31	30	12	31	223
Lahore	109 *74*	20 *20*	31 *31*	50 *35*	78	26	674
Manşurah	105	0	27	40	82	50	576

[1] See page 25: This is a selection only. The small figures in italics are approximate modern values. For lists see Ptolemy, al-Battānī, Ulugh Beg., the *Āin-i-akbari*, etc.

[2] Jerusalem.

[3] 119 on 'Jaipur D.'

[4] The values vary from 113° 0′ to 113° 35′ and from 28° 15′ to 29° 0′.

[5] Aleppo.

[6] Alexandria.

C. (1) Astrolabe Gazetteer—*contd.*

Place.	Longitude. °	′	Latitude. °	′	Inhirāf. °	′	Distance.
Marāghah	82 *46*	0 *17*	37 *37*	20 *21*	16	17	360
Miṣr[1]	63 *31*	20 *15*	30 *30*	20 *2*	58	38	335
Moṣul	77 *43*	0 *9*	34 *36*	30 *19*	4	12	285
Nishābūr	92 *58*	30 *10*	36 *36*	21 *8*	46	25	440
Qāīn	93	20	33	40	54	1	414
Qandahar	107 *65*	40 *40*	33 *31*	0 *36*	75	0	656
Qazwīn	85	0	35	0	27	34	352
Sabzawār	91	30	36	5	44	12	422
Samarqand	99 *66*	26 *59*	39 *34*	37 *40*	..		..
Shirāz	88 *52*	0 *40*	29 *29*	36 *30*	53	58	279
Sirhind	111 *76*	33 *23*	30 *30*	30 *37*	..		..
Ṭarablis[2]	45 *12*	0 *26*	32 *33*	0 *49*	78	17	674
Tabriz	82 *46*	0 *12*	38 *38*	0 *2*	15	40	367
Tiflis	83 *44*	0 *50*	43 *41*	0 *35*	14	41	486
Ujjain	102[3] *75*	0 *47*	24 *23*	30 *10*	77	57	510
Yezd	89 *54*	0 *30*	32 *31*	0 *55*	48	29	331

[1] Cairo.
[2] Tripoli.
[3] Possibly wrongly copied for 112 : I_2 gives 110° 50′ and 23° 30′.

C. (2) Longitudes and Latitudes of Ujain, Delhi, Benares, Jaipur.

Name of Place.	Date.	Authority.	Locality of observation.	Longitude.	Latitude North.
	A. D.				
Ujjain	550	*Pañchasiddhāntikā*	....	44° East of Alexandria.[1]	24° 0′ 0″
	1000	Hindu canon according to Alberuni.	....	....	24° 0′ 0″
	1010	Alberuni	....	....	22° 49′ 0″
	1150	Bhāskara	....	....	22° 30′ 0″
		Astrolabes	....	110° 50′ East of Fortunate Isles.[2]	23° 30′ 0″

[1] The longitude of modern Alexandria is given as 29° 51′ East of Greenwich. The astrolabes general gave 61° 54′ East of the Fortunate Isles (see p. 127).
[2] For the longitude of the Fortunate Isles see p. 25 where it is estimated at 35° West of Greenwich.

C. (2) Longitude and Latitudes—*contd.*

Name of Place.	Date.	Authority.	Locality of observation.	Longitude.	Latitude North.
UJJAIN—*contd.*	C. 1734	Jai Singh . .	Observatory . .	....	23° 10′ 0″.
	1792	W. Hunter (*As. Res.* 1795, p. 141f).	Near Rana Khan's Garden.	75° 46′ East of Greenwich.	23° 11′ 34·2″.
		W. Hunter . .	Near Scindia's Palace	75° 55′ 3″ East of Greenwich.	23° 10′ 56·5″.
	1825	Warren from Hindu Books.	....	....	23° 11′ 30″.
	1915	Trigonometrical Survey of India.	Hill 1678 . .	....	23° 11′ 6″.
DELHI . . .	..	Astrolabes . .	....	113° 15′ East of Fortunate Isles.[1]	28° 39′ 0″.
	1729	Jai Singh . .	Observatory . .	....	28° 39′ 0″.
	1734	Father Boudier .	" . .	75° 0′ East of Paris[2]	28° 37′ 0″.
	1792	W. Hunter . .	" . .	77° 2′ 27″ East of Greenwich.	28° 37′ 0″.
	1825	Warren from Hindu Books.	....	1° 16′ 8″ East of Ujjain.	27° 35′ 0″.
	1915	Trigonometrical Survey of India.	Pirghaib . . Jama Masjid . .	77° 12′ 52″ . . 77° 14′ 2·8″ . .	28′ 40′ 35·1″. 28′ 39′ 2·3″.
BENARES . .	550	*Pañchasiddhāntikā* .	....	54° East of Alexandria.	
	..	Astrolabes . .	....	117° 20′ East of Fortunate Isles.	26° 15′ 0″.
	1795	R. Barrow *As. Res.*, IV 1795, p. 326.	....	82° 54′ East of Greenwich.	25° 18′ 36″.
	1825	Warren from Hindu Books.	....	4° 37′ 0″ East of Ujjain.	25° 38′ 0″.
	1915	Trigonometrical Survey of India.	Observatory . .	83° 0′ 46·1″ East of Greenwich.	25° 18′ 24·9″
JAIPUR . . .	..	Astrolabe . .	....	109° 6′ East of Fortunate Isles.	26° 36′.
		Tieffenthaler . .	....	73° 43′ East of Paris	26° 53′.
	1734	Father Boudier .	Observatory . .	73° 50′ East of Paris	26° 56′.
	1915	Trigonometrical Survey of India.	Minaret. . .	75° 49′ 18·5″ East of Greenwich.	26° 55′ 27·4″.

[1] For the longitude of the Fortunate Isles see p. 25 where it is estimated at 35° West of Greenwich.
[2] The longitude of Paris Observatory is 2° 20′ 13·5″ East of Greenwich.

C. (3) Observatory Elements.

(a) *Position.*

	Latitude.	Longitude East of Greenwich.		Magnetic Declination for 1915.	Annual Variation.
			h. m. s.	East.	
DELHI	28° 37′ 35″	77° 13′ 5″	5 8 52·3	1° 45′	—1
JAIPUR	26° 55′ 27·4″	75° 49′ 18″	5 3 17·2	1° 45′	
UJJAIN	23° 10′ 6″	75° 4[illegible]′ [illegible]″ 46 2	5 3 [illegible] 4	0° 45′	
BENARES	25° 18′ 24·9″	83° 0′ 46″	5 32 3·1	1° 5′	

SIMLA

(b) *Time.*

Place.	Longitude East of Greenwich.	Difference between local and standard time.	DIFFERENCE BETWEEN LOCAL SUN TIME AND CLOCK (i.e., STANDARD) TIME. Jan. 1.	Feb. 12.	Mar. 15.	April 15.	May 15.	June. 14.	July 15.	Aug. 15.	Sept. 1.	Oct. 1.	Nov. 3.	Dec. 25
		M. S.	M. S.	M. S.	M. S.	M. S.	M. S.	M. S.	M. S.	M. S.	M. S.	M. S.	M. S.	M. S.
Standard . . .	82° 30′ 0″	0 0	+3 11	+14 24	+9 6	0 0	—3 48	0 0	+5 44	+4 23	0 0	—10 16	—16 21	0 0
Delhi . . .	77° 13′ 5″	+21 8	+24 1	+35 32	+30 14	+21 8	+17 20	+21 8	+26 52	+25 31	+21 8	+10 52	+4 47	+21 8
Jaipur . . .	75° 49′ 18″	+26 43	+29 54	+41 7	+35 49	+26 43	+22 55	+26 43	+32 27	+31 6	+26 43	+16 27	+10 22	+26 43
Ujjain . . .	75° [illegible] 2	+26 52	+30 3	+41 16	+35 8	+26 52	+23 4	+26 52	+32 36	+31 15	+26 52	+16 36	+10 31	+26 52
Benares . . .	83° 0′ 46″	—2 3	+1 8	+12 21	+7 3	—2 3	—5 51	—2 3	+3 41	+2 20	—2 3	—12 19	—18 24	—2 3

Standard time in India is 5½ hours before Greenwich time. The minus sign indicates that standard time is behind local sun time. Therefore to find standard time from dial time add the *plus* quantities and subtract the *minus* quantities as given in the table. For intermediate dates proportionate parts will give approximate result. The time is given to the nearest second only for 1916. The annual variation is small.

C. (4) Climates and Longest days.

Climates.		Hours of longest day.	Latitudes.				Longest Days.	
			After Ptolemy.	Al-Battānī.[1]	On Jaipur A Astro-labe.	Calculated from hours	Calculated from latitudes.	
							Latitudes.	H. M.
First	b	12¾	12° 30′	..	12° 43′	..	20	13 13
	m	13	16° 27′	16° 39′	16° 44′	16° 43′	21	13 17
Second	b	13¼	20° 14′	20° 28′	20° 31′	..	22	13 21
	m	13½	23° 51′	24° 5′	24° 10′	24° 10′	23	13 25
Third	b	13¾	27° 12′	27° 28′	27° 34′	..	24	13 29
	m	14	30° 2′	30° 40′	30° 46′	30° 46′	25	13 34
Fourth	b	14¼	33° 18′	33° 37′	33° 43′	..	26	13 38
	m	14½	36° 0′	36° 22′	36° 28′	36° 28′	27	13 42
Fifth	b	14¾	38° 35′	38° 54′	39° 1′	..	28	13 47
	m	15	40° 56′	41° 15′	41° 21′	41° 21′	29	13 52
Sixth	b	15¼	43° 41′	43° 25′	43° 30′	..	30	13 56
	m	15½	45° 1′	45° 22′	45° 39′	45° 32′	40	14 51
Seventh	b	15¾	46° 51′	47° 12′	47° 38′	..	50	16 10
	m	16	48° 32′	48° 53′	48° 59′	48′ 59°	60	18 31
		16¼	50° 4′ etc.	..	..	..	..	..

[1] *Opus Astronomicum* Ed. Nallino 2nd part 65—66.

The calculated results are obtained from the formula $\frac{180^\circ + 2h}{15}$ = *the longest day* where *sin h = tan φ. tan ω* and where *ω* = 23° 29′ and *φ = the latitude.*

APPENDIX D.

Technical Terms and Tables.

D. TECHNICAL TERMS AND SYMBOLS.

D (1). Numerical Notations.

	Ajad.		Arabic Numerals.	Hindu Numerals.		Abjad.		Arabic Numerals.	Hindu Numerals.
1	ا	a	١	१	60	س	s	٦٠	६०
2	ب	b	٢	२	70	ع	'	٧٠	७०
3	ج	j	٣	३	80	ف	f	٨٠	८०
4	د	d	٤	४	90	ص	ṣ	٩٠	९०
5	ه	h	٥	५	100	ق	q	١٠٠	१००
6	و	w	٦	६	200	ر	r	٢٠٠	२००
7	ز	z	٧	७	300	ش	sh	٣٠٠	३००
8	ح	ḥ	٨	८	400	ت	t	٤٠٠	४००
9	ط	ṭ	٩	९	500	ث	th	٥٠٠	५००
10	ي	i	١٠	१०	600	خ	kh	٦٠٠	६००
20	ك	k	٢٠	२०	700	ذ	dh	٧٠٠	७००
30	ل	l	٣٠	३०	800	ض	ḍ	٨٠٠	८००
40	م	m	٤٠	४०	900	ظ	ẓ	٩٠٠	९००
50	ن	n	٥٠	५०	1000	غ	gh	١٠٠٠	१०००

The *abjad* system is indicated by eight *voces memoriabiles*
abjad hawwaz ḥuṭṭi kalaman sa'faṣ qarashat thakhadh daẓagh.

But western Mussulmans arrange the last four words thus : *Safad qarasat thakhadh ẓaghash.*

The Greek system agreed with the abjad as far as 80 but the portion from 90 to 300 corresponded with the *abjad* from 100 to 400. The later numbers differed considerably. In deciphering the engraved numbers on the astrolabe there are sometimes apparent ambiguities to be cleared up, owing, partly, to the omission of diacritical marks.[1]

[1] See page 97 ; Nallino's *al-Battānī*, *iii f.* and Peters and Knobel's *Ptolemy's Catalogue of Stars* p. 13 *f.* Morely made a number of mistakes particularly in transcribing the symbols for 4 and 7.

D (2). Signs of the Zodiac.

					Arabic.	Sanskrit.
	0	♈	♈	Aries	al-Ḥamal . .	Meshā.
30	1	♉	♉	Taurus	al-Thaur . .	Vṛishan.
60	2	♊	♊	Gemini	al-Jauzā . .	Mithuna.
90	3	♋	♋	Cancer	al-Saraṭān . .	Karka.
120	4	♌	♌	Leo	al-Asad . .	Sinha.
150	5	♍	♍	Virgo	al-Sumbulah . .	Kanyā.
180	6	♎	♎	Libra	al-Mīzān . .	Tulā.
210	7	♏	♏	Scorpio	al-'Aqrab . .	Vṛiśchika.
240	8	♐	♐	Sagittarius	al-Qaus . .	Dhanus.
270	9	♑	♑	Capricornus . . .	al-Jadī . . .	Makara.
300	10		♒	Aquarius	al-Dalau . .	Kumbha.
330	11		♓	Pisces	al-Hūt . . .	Mina.

D (3). The Planets.

			Arabic.[1]	Sanskrit.[2]
1	♄	Saturn	Zuḥal . . .	Ārki.
2	♃	Jupiter	Mushtārī . .	Bṛihaspati.
3	♂	Mars	Mirrīkh . . .	Kuja.
4	☉	Sun	Shams . . .	Sūrya.
5	♀	Venus	Zuhrah . . .	Śukra.
6	☿	Mercury	'Uṭarid . . .	Budha.
7	☽	Moon	Qamar . . .	Chandra.

[1] The planets with the exception of the sun are sometimes (*e.g.*, in the table of Trigons on Jaipur A) denoted by the final letters of the Arabic names *l*, *ī*, *kh*, (*sh*), *h*, *d*, *r*; See also Sédillot *Prol. Tables Astron. d'Oulug B. g* p. cxlviii.

[2] There are many variants of these names.

D (4). Lunar Mansions.

Hindu Nakshatras.	Arabic Manzils.
1. Aśvini (β, γ *Arietis*)	1. Sharaṭin (β, γ *Arietis*).
2. Bharaṇi (35, 39, 41 *Arietis*)	2. Buṭin (35, 39, 41 *Arietis*).
3. Kṛittikā (*Pleiades*, η *Tauri*, *&c.*)	3. Thuraiya (*Pleiades*, η *Tauri*, *&c.*)
4. Rohini (α, θ, γ, δ, ε *Tauri*)	4. Dabarān (α, θ, γ, δ, ε *Tauri*).
5. Mṛigaśiras (λ, φ, $φ_2$ *Orionis*)	5. Haqāh (λ, φ, $φ_2$ *Orionis*).
6. Ārdrā (α *Orionis*)	6. Han'ah (η, μ, ν, γ, ξ *Geminorum*).
7. Punarvasu (β, α *Geminorum*)	7. Dhirā (β, α *Geminorum*).
8. Pushya (θ, δ, γ *Cancri*)	8. Nathrah (γ, δ *Cancri*).
9. Aśleshā (ε, δ, σ, η, ρ *Hydræ*)	9. Ṭarf (ξ *Cancri*, λ *Leonis*).
10. Maghā (α, η, γ, ζ, μ, ε *Leonis*)	10. Jabhah (α, η, γ, ζ *Leonis*).
11. Pūrva-Phalgunī (δ, θ *Leonis*)	11. Zubrah (δ, θ *Leonis*).
12. Uttara-Phalgunī (β, 93 *Leonis*)	12. Sarfah (β *Leonis*).
13. Hasta (δ, γ, ε, α, β *Corvi*)	13. Auwā (β, η, γ, δ, ε *Virginis*).
14. Chitrā (α *Virginis*)	14. Simāk (α *Virginis*).
15. Svāti (α *Bootis*)	15. Ghafr (ι, κ, λ *Virginis*).
16. Viśākhā (ι, γ, β, α *Libræ*)	16. Zubānān (α, β *Librae*).
17. Anurādhā (δ, β, π, *Scorpii*)	17. Iklil (β, δ, π *Scorpii*).
18. Jyeshṭhā (α, σ, τ *Scorpii*)	18. Qalb (α *Scorpii*).
19. Mūla (λ, υ, κ, ι, θ, η, ζ, μ, ε *Scorpii*)	19. Shaulah (λ, υ *Scorpii*).
20. Pūrva-Ashāḍhā (σ, ε *Sagittarii*).	20. Na'āim ($γ^2$ — ζ *Sagittarii*).
21. Uttara-Ashāḍhā (δ, ζ *Sagittarii*)	21. Baldah (*North of* π *Sagittarii*).
22. Abhijit† (α, ε, ζ *Lyræ*)	22 Sa'd al-Dhābih (α, β *Capricorni*).
23. Śravana (α β γ *Aquilæ*)	23. Sa'd Bula' (ε, μ, ν *Aquarii*).
24. Śravishtha (β, α, γ, δ *Delphini*)	24. Sa'd al-Su'ud (β, ζ *Aquarii*).
25. Śatabishaj (λ *Aquarii*, *etc.*)	25. Sa'd al-Akhbiyah (α, γ, ζ, η *Aquarii*).
26. Pūrva-Bhādrapadā (α, β *Pegasi*)	26. al-Fargh al-Mugdim (α, β *Pegasi*).
27. Uttara-Bhādrapadā (γ *Pegasi*, α *Andromedæ*)	27. al-Fargh al-Mukhir (γ *Pegasi*, α *Andromedae*).
28. Revati (ζ *Piscium*, *etc.*)	28. Batn al-Hut. (β *Andromedæ*, *etc.*)

D (5). Obliquity of the Ecliptic.

B. C. 130	Hipparchus	23° 51′ 20″.
A. D. 150	Ptolemy	23° 51′ 15″.
880	al-Battānī	23° 35′ 0″.
965	Abul Raḥman al-Sufi	23° 33′ 45″.
1001	Ibn Yūnus	23° 34′ 52″.
1100	*Sūrya Siddhānta*	24° 0′ 0″.
1250	Alphonsine tables	23° 32′ 39″.
1270	Naṣir al-Din al-Ṭūsi	23° 30′ 0″.
1438	Ulugh Beg	23° 30′ 17″.
1590	Tycho Brahe	23° 29′ 47″.
1690	Flamsteed	23° 28′ 48″.
1729	Jai Singh	23° 28′ 0″.
1900	....	23° 27′ 28″.
1916	1st July	23° 27′ 4·68″.

(The secular variation is between 21° 59′ and 24° 36′ approximately.)

D (6). Length of the year.

	D.	H.	M.	S.	
Hipparchus	365	5	55	12	Tropical.
Ptolemy	365	5	46	24	"
"	365	6	9	48·6	Sidereal.
Romaka Siddhānta	365	5	55	12	
Pauliśa Siddhānta	365	6	12	36	Sidereal.
Āryabhaṭa	365	6	12	30	"
Brahma Gupta	365	6	12	30·915	"
al-Battānī	365	6	12	—	"
Siddhānta Śiromaṇi	365	6	12	9	"
Astrolabe	365	5	50	12·4	Tropical.
Approximately correct values—					
Tropical year	365	5	48	45·98.	
Sidereal year	365	6	9	9·5.	

D (7). Precession of the Equinoxes.

Annual precession.

A. D.		
150	Ptolemy.	36″ a year or 1° in 100 years.
810	al-Battānī	54·55″ a year or 1° in 66 years.
1100	*Sūrya Siddhānta*	54″ a year.
1150	Bhāskara	59·9″ a year.
1370	Mahendra Sūri	54″ a year.
1500	Copernicus	50·2″ a year.
1590	Tycho Brahe	51″ "
1729	Jai Singh	51·6″ or 4° 8′ in 297 Muslim years.

[The precession for celestial longitude of all stars is approximately 50° 26″ a year or, more correctly 50·2564″+0·0222″ T (where T is the time in centuries reckoned from A.D. 1900). The complete period of precession is approximately 25,696 years.]

APPENDIX E.

Chronology

E. CHRONOLOGY.

There has been considerable difficulty in ascertaining the dates of construction of the observatories, etc. Sir Robert Barker, who was contemporary with Jai Singh, stated that the observatory at Benares was supposed to have been built by Akbar; James Prinsep gave the date as A.D. 1680—six years before the birth of Jai Singh; another writer gives 1693 and so on: of the following dates those relating to Jai Singh must consequently be used with some circumspection.

	A.D.
Greenwich observatory founded	1675
Jai Singh born	1686
Newton's *Principia* published	1687
Jai Singh succeeds to the Amber Gadi	1699
Halley's *Synopsis of Cometary Astronomy*	1705
Death of Aurangzib	1707
Jai Singh invests Thun	1712
G. D. Cassini dies	1712
Farruk-Siyar	1713
Jai Singh displaces Budh Singh of Būndi	1718
Muhammad Shāh	1719
Jai Singh appointed the Emperor's deputy at Agra and Mālwa	1719
Great earth-quake at Delhi	1720
Flamsteed dies	1720
Jai Singh's expedition against Jāts of Bharatpur	1722
Delhi observatory built	C. 1724
Historia Cœlestis Britannica published	1725
Isaac Newton died	1727
Jaipur city built	C. 1728
The *Zij Muhammad Shāhī* finished	C. 1728
Aberration of light discovered by Bradley	1728
Observations recorded at Delhi	1729
Jai Singh resigns the province of Mālwa to the Peshwā	1734
Jaipur observatory built	1734
Benares observatory built	1737
Nādir Shāh sacked Delhi	1739
Halley died	1742
Jai Singh died	1743

APPENDIX F.

Bibliography.

F. BIBLIOGRAPHY.

(1) Jai Singh's Observatories.

SIR R. BARKER.—An account of the Bramins' observatory at Benares. By Sir Robert Barker. In a letter to Sir John Pringle, P.R.S. (*Philosophical Transaction of the Royal Society of London.* Vol. LXVII, 1777, part ii, No. XXX, p. 598 f.)

J. L WILLIAMS.—Further particulars respecting the observatory at Benares. (*Philosophical Transactions of the Royal Society of London.* 1793, part i, p. 45 f.)

W. HUNTER.—Some Account of the astronomical labours of Jayasinha, Rajah of Ambhere or Jayanagar. (*Asiatic Researches or Transaction of the Society instituted in Bengal.* Vol v, 1799, p. 177f. and p. 444.)

Lettres édifiantes et curieuses, écrites des Missions étrangères. Nouvelle edition. Mémoire des Indes. Tome quinzième. Toulouse, 1810.

J. TOD.—Annals and antiquities of Rajast'han, or the Central and Western Rajpoot States of India. (Vol. ii, p. 345f. Annals of Amber, or Dhoondar) *London,* 1832.

GARCIN DE TASSY.—Description des monuments de Delhi en 1852. D'après le texte Hindu-oustani de Saiyid Ahmed Khan. (*Journal asiatique.* Cinquième serie. Tome xvi. Decembre 1860, pp. 536-543.)

A. ff. GARRETT AND CHANDRADHAR GULERI.—The Jaipur observatory and its Builder. *Allahabad,* 1902.

SEVERIN NOTI, S. J.—Land und Volk des königlichen Astronomer Dschaisingh II Maharadscha von Dschaipur. *Berlin,* 1911.

See also D'ANVILLE *Eclaircissemens géographiques,* 1753, 46f. and *Ant. Géog. de l'Inde.* 1775, 6 f.; J. BERNOULLI, Des Pater Joseph Tieffenthalers historisch-geographische Beschreibung von Hindustan, 1785, i, 143. 224. 246; RENNELL, *Memoir of a map of Hindoostan* 3rd Ed. 1793, 78; *Journ. Arch. Socy.* Delhi, 1853; SIR J. HOOKER Him. Journ. 1854, 74-77; BAPUDEVA ŚĀSTRI, The Mānmandera Observatory (Pamphlet, 6 pp.), 1865; HENDLEY, *Hand-Book Jeypore Courts,* 1886; J. BURGESS, *Indian Antiquary,* 1906, XXXV, 234-235; LALA CHIMANLAL The Delhi and Benares observatories (Pamphlet, 14 pp.), 1912; H. G. TOMKINS, *Journ. Cal. Astr. Socy.,* 1912, 71-76.

(2) Astrolabes.

J. J. MIDDLETON.—Description of a Persian Astrolabe, submitted to the Asiatic Society by Major Pottinger. (*Journal of the Asiatic Society of Bengal,* 1841, x, 761 f.)

W. H. MORLEY.—Description of a planispheric astrolabe constructed for Shah Sultán Hussain Safawi, King of Persia, and now preserved in the British Museum. *London.*

H. SAUVAIRE AND J. DE REY PAILHADE.—Sur une Mère d'Astrolabe arabe du xiiie siècle, etc. (*Journal Asiatique,* Neuvième série, Tome i, 1893, pp. 5-76 and pp. 185-231.)

M. L HUGGINS.—The Astrolabe—A summary. (*Astronomy and Astrophysics,* vol. xiii, 1894, pp. 793-802.) (See also article 'Astrolabe' in the *Encyclopædia Britannica.*)

CARRA de VAUX.—L Astrolabe linéaire ou Baton d'et-Tousi. (*Journal asiatique,* Nouvième série, Tome V, Mai-Juin, 1895, pp. 464-516.)

NAU F.—Le traité sur l'Astrolabe plan de Sévère Sabokt. *Jour. asiat.* 1899, 56-101, 238-303.

H. S. COWPER.—A Fifteenth Century Planispheric Astrolabe, made at Granada. (*Journa of the Royal Asiatic Society of Great Britain and Ireland,* 1904, pp. 53-66 and 542-3.)

S. A. IONIDES.—Description of an astrolabe. (*The Geographical Journal*, 1904, xxiv, pp. 411-417.)

C. A. NALLINO.—Asṭurlāb (*Encyclopædia of Islam*, 1913, f.).

See also DELAMBRE, SÉDILLOT, etc., below; also FOCARD, *Paraphrase, de l'Astrolabe* 1555; CLAVIUS, *Gnomonices*, 1581, and *Astrolabium*, 1593, and list in MORLEY'S *Plan. Astr.*, p. 46. For Arabic-Persian works on the astrolabe (which are exceedingly numerous), see SUTER'S, *Math. u. Astr. der Araber*. Persian lithographed text-books are fairly common. See also PRATT, *J. A. S. B.*, 1842, xi, 720-721; MORGAN, *Archæologia*, 1851, xxx, 259-264; LECKY *Monthly Notices*, R. A. S., xxxiv, 120; *also* CHAUCER'S *Treatise on the Astrolabe* (A.D. 1691) and SKEAT'S and BRAE'S notes thereon; L. EVANS *Arch. Jour.* 1911, 221 f.

(3) Hindu Astronomy.

(a) Vedic.

B. G. TILAK.—The Orion, or Researches into the antiquity of the Vedas. *Bombay*, 1893.

W. D. WHITNEY.—On a recent attempt by Jacobi and Tilak to determine on astronomical evidence the date of the earliest Vedic period as 4000 B.C. (*Proceedings of the American Oriental Society*, 1894, and the *Indian Antiquary*, xxiv, 1895, pp. 361-369.)

G. THIBAUT.—On some recent attempts to determine the antiquity of Vedic civilization. (*Indian Antiquary*, xxiv, 1895, pp. 85-100.)

See also A. WEBER, *Die vedeschen nachrichten vom den Naxatra* 1860—1862, and *Uber den Veda Kalendar*, 1862, and *Ind. Stud.* x, 1868, 213-253; B. G. TILAK, *Trans. Ninth Internal. Cong. Orientalists*, 376-383; H. JACOBI, *Festgruss an Roth*, 1893, 68-74; *Ind. Ant.* xxiii, 1894 154f., *Z. D. M. G.*, XLIX, 1895, 218-230, and L, 1896, 69-83; G. BÜHLER, *Ind. Ant.* xxiii, 1894, 238-249; H. OLDENBERG, *Z. D. M. G.*, XLVIII, 1894, 629-648, XLIX, 1895, 470-478. DIKSHIT, *Ind. Ant.*, 1895, xxiv, 245-6; FLEET, *J.R.A.S.*, 1911, 514-518 and 1119; BERRIEDALE KEITH, *J.R.A.S.*, 1911, 794 and 1914, 627-640; R. SHAMASASTRY, *Ind. Ant.*, xli, 1912, 26, 45, 77, 117, 290 and xlii, 1913, 19, 32, 72; N. K. DUTT, *Dacca Review*, 1915, v, 270 ff, etc.

(b) Texts and Commentaries.

[W. D. WHITNEY AND] E. BURGESS.—Translation of the Sūrya Siddhānta, a Text-book of Hindu Astronomy, with notes and appendix by Rev. Ebenezer Burgess, assisted by the committee of Publication. (*Journal of the American Oriental Society*, 1858, pp. 141-498.)

BAPU DEVA SASTRI.—Translation of the Sūrya Siddhānta and of the Siddhānta Śiromaṇi by the late Lancelot Wilkinson, revised by B. D. Shastri, *Calcutta*, 1861.

*H. KERN.—The Āryabhatīyas with the Commentary, Batadipika of Paramadisvara. *Leiden*, 1874.

G. THIBAUT AND SUDHAKARA DVIVEDI.—The Pañchasiddhāntikā. The astronomical work of Varāha Mihira. The text edited with an original commentary in Sanskrit and an English Translation and Introduction. *Benares*, 1889

*BRĀHMASPHUṬASIDDHANTA AND DHYANAGRAHOPADESHADHYAYA BY BRAHMAGUPTA.—Edited with his own commentary by Mahāmahopādhyaya Sudhākara Dvivedin. *Benares*, 1902.

LALA CHHOTE LAL.—The obscure text of the Jyotisha Vedāṅga explained. *Allahabad*, 1907.

*BHĀSKARA.—The Treatise on astronomy by Bhāskarāchārya with his own exposition, the Vasana Bhashya. Formerly edited by the late Mahamahopadhyaya Pandit Bapu Deo Shastri.

See also ROER, *J. A. S. B.*, 1844; A. WEBER, *Ind. Stud.*, x, 1866, 213 f.; HOERNLE, *J. A. S.B.*, 1893, LVII, 9-18; THIBAUT, *J. A. S. B.*, 1877, XLVI, 411-437, 1880, XLIX, 107-127

* In Sanskrit only.

and 181-206 and 1884 LIII, 259-293 ; J. BURGESS *Ind. Ant.*, 1891, xx, 53 ; DIKSHIT *Ind. Ant.* XIX, 1890 ; SUDHĀKARA DVIVEDĪ's, *Jyantisham*, Benares, 1908 ; xix, 45-54 ; M. P. KAREGAT, *J. A. S. Bombay*, 1896 ; *The Benares Sanskrit Series*, etc.

(c) *General.*

J. S. BAILLY.—Traité de l'Astronomie Indienne et Orientale. 1787.

J. BENTLEY.—A Historical view of Hindu Astronomy from the earliest dawn of that science in India to the present time. In two Parts, part i. The Ancient Astronomy. Part ii The Modern Astronomy, etc. *London*, 1825.

J. M. F. GUERIN.—Astronomie Indienne d'après la Doctrine et les livres anciennes et modernes des Brammes sur l'Astronomie, l'Astrologie et la Chronolgie—Suivie de l'examen de l'astronomie des anciens peuples de l'Orient et de l'Explication des principaux Monuments astronomico-astrologiques de l'Egypte et de la Perse. *Paris*, 1847.

* SUDHĀKARA DVIVEDI.—Gaṇaka Taraṅgiṇī, or Lives of Hindu Astronomers.

J. BURGESS.—Notes on Hindu astronomy and the History of our knowledge of it. J.R.A.S 1893, pp. 717 f.

* S. B. DIKSHIT.—Bhāratīya Jyotih-Śāstra. *Poona*, 1896.

G. THIBAUT.—Astronomie, Astrologie und Mathematik (*Grundriss der indo-arischen Philologie*, iii, 9, 1897), pp. 717 f.

See also DAVIS, *As. Res.*, 1790, ii, 225-287 and 1792, iii, 209-227 ; JONES, *As. Res.*, 1790, ii, 289-306, 1792, iii, 257-294 ; J. PLAYFAIR, *As. Res.*, 1795, iv, 158-164 ; J. BENTLEY, *As. Res.*, 1799, vi, 537-588 ; BUCHANAN, *As. Res.*, 1799, vi, 163 f. ; JONES, *Works*, iv, 1-47, 48-69, 71-92 ; J. BENTLEY, *As. Res.*, 1805, viii, 193-244 ; H. T. COLEBROOKE, *As. Res.*, 1807, ix, 323-376 and 1816, XII, 209-250 [also *Misc. Essays*, 1872, II, xiv, 321-373 and II, xv, 374-416] ; T. WARREN, *Kala Sankalita* 1825 ; H. R. HOISINGTON, *Oriental Astronomer*, Jaffna, 1848 ; W. BRENNAND, *Hindu Astronomy*, 1896 ; T. AUFRECHT, *Cat. Bod.*, viii, 325 f., 1864 ; J. EGGELING, *Cat. Sans. MSS.*, India office, part V, x, 1896, *Calcutta Review*, 1850, xxv, 65-85 ; 1900, cx, 344-356 ; FLEET, *J. R. A. S.*, 1915, 213-230, etc., etc. A. BOUTQIN, *Ciel et Terre*, Bruxelles, 1908, pp. 105-112, 129-133, 159-163, 213-222.

(d) *Chronology.*

See DIKSHIT, *Corp. Inscr. Ind.*, 1888, iii, 161-176 ; JACOBI, *Ind. Ant.*, 1888, xvii, 145-181, *Ep. Ind.*, i, 403-460 and ii, 487-498 ; Schram, *Ind. Ant.*, 1889, xviii, 290-300 ; DIKSHIT, *Ind. Ant.*, 1890, xix, 48 f. ; SEWELL and DIKSHIT, *The Indian Calendar*, 1896 ; KIELHORN, *Ind. Ant.*, 1899, xviii, 193-209 and 380-386 ; R. SCHRAM, *Kalendariographische und Chronologische Tafeln*. 1908 ; FLEET, *Enc. Britannica*, 13, 491-501 ; PILLAI's, *Indian Chronology*, 1911 ; SEWELL, *Indian Chronography*, 1912 ; etc.

(4) Muslim Astronomy.

DELAMBRE.—Histoire de l'astronomie du Moyen Age. *Paris*, 1819.

J. J. SÉDILLOT.—Traité des instruments astronomiques des Arabes. Traduit d'Aboul-Hassan par J. J. Sedillot, et publié avec une introduction. 2 vols., 1834-35.

L. A. SÉDILLOT.—Mémoire sur les instruments des Arabes. Paris, 1845.

L. P. E. A. SÉDILLOT.—Prolégomènes des Tables astronomiques d'Oloug-Beg. Publiés avec notes et variantes et précédés d'une Introduction. *Paris*, 1847.

L. P. E. A. SÉDILLOT.—Prolégomènes des Tables astronomiques d'Oloug-Beg Traduction et commentaire. *Paris*. 1853.

C. A. NALLINO.—Al Battānī sive Albatenii. Opus Astronomicum. Milan, 1899-1907.
H. SUTER.—Die Mathematiker und Astronomer der Araber und Ihre Werke. *Leipzig*, 1900,
G. A. NALLINO.—Astronomy (*Encyclopædia of Islām*, 1913, f.).

See also JOURDAIN, Mémoire sur l'observétoire de Méragah, 1810; CARRA DE VAUX, l'Almageste d'Abu'l Wefa, etc., *Jour. As.*, 1892, XIX, 408-471; L. A. HOCHEIM, *Z.D.M.G.*, 1893, XLVII, 213-275; and FLAMSTEED'S *Hist. Cœl. Brit.* for Ulugh Beg's Catalogue, etc, etc.

(5) Astrology.

H. KERN—Bṛihat-Saṁhita. J. R. A. S. 1870-75 and *Verspereide Geschriften*, i & ii. 1893.
A. BOUCHÉ-LECLERCQ—L'Astrologie grecque. *Paris*, 1899.
HARI PRASANNA CHATTERJEE.—The Brihājjātakam of Varāha Mihira, *Allahabad*, 1912.
C. A. NALLINO.—'Astrology' (*Encyclopædia of Islām*).

See also A. WEBER, *Ind. Stud.*, ii, 1853, 236-315 and iii; H. JACOBI, *Indische Streifen*, iii, 1879, 165-168, etc.

(6) General.

J. FLAMSTEED.—Historia Cœlestis Britannica. 1725.
DELAMBRE.—Histoire de l'astronomie ancienne. *Paris*, 1817.
DELAMBRE.—Histoire de l'astronomie du moyen age. *Paris*, 1819.
F. BAILY.—The Catalogues of Ptolemy, Ulugh Beigh, Tycho Brahé, Halley, Hevelius *Memoirs Roy. Astr. Soc.* 1843, vol. xiii.
R. WOLF.—Geschichte der Astronomie. *München*, 1877.
E. B. KNOBEL.—The Chronology of Star Catalogues. *London*, 1877.
J. L. E. DREYER.—Tycho Brahe: a picture of scientific life and work in the sixteenth century, *Edinburgh*, 1890.
A. BERRY.—A short History of Astronomy. *London*, 1898.
A. VON BRAUNMÜHL.—Vorlesungen über Geschichte der Trigonometrie. *Leipzig*, 1900.
J. L. E. DREYER.—History of the Planetary systems from Thales to Kepler. *Cambridge*, 1906.
G. FORBES.—History of Astronomy.[1] *London*, 1909.
C. H. E. PETERS AND E. B. KNOBEL.—Ptolemy's Catalogue of Stars. A Revision of the Almagest. *Washington*, 1915.

[1] A cheap and excellent little book.

INDEX

(The references are to the pages.)

A

B